Mathematical Explorations with MATLAB

This book is about the kind of mathematics usually encountered in
first year university courses. A key feature of the book is that this
mathematics is explored in depth using the popular and powerful
package MATLAB. The emphasis is on understanding and investigating
the mathematics, and putting it into practice in a wide variety of
modelling situations. In the process, the reader will gain some fluency
with MATLAB, no starting knowledge of the package being assumed.
The range of material is wide: matrices, whole numbers, complex
numbers, geometry of curves and families of lines, data analysis,
random numbers and simulations, and differential equations form the
basic mathematics. This is applied to a large number of investigations
and modelling problems, from sequences of real numbers to cafeteria
queues, from card shuffling to models of fish growth. All extras to the
standard MATLAB package are supplied on the World Wide Web.

All three authors hold positions at the University of Liverpool.
Ke Chen is Lecturer in Mathematical Sciences, Peter Giblin is Reader
in Mathematics, and Alan Irving is Reader in Theoretical Physics.

Mathematical Explorations with MATLAB

KE CHEN, PETER GIBLIN, ALAN IRVING

CAMBRIDGE
UNIVERSITY PRESS

PUBLISHED BY THE PRESS SYNDICATE OF THE UNIVERSITY OF CAMBRIDGE
The Pitt Building, Trumpington Street, Cambridge, United Kingdom

CAMBRIDGE UNIVERSITY PRESS
The Edinburgh Building, Cambridge, CB2 2RU, UK   http://www.cup.cam.ac.uk
40 West 20th Street, New York, NY 10011-4211, USA   http://www.cup.org
10 Stamford Road, Oakleigh, Melbourne 3166, Australia

First published 1999

Printed in the United Kingdom at the University Press, Cambridge

Typeset in Computer Modern 10/13pt by the authors using LaTeX2e

A catalogue record of this book is available from the British Library

ISBN 0 521 63078 9 hardback
ISBN 0 521 63920 4 paperback

Contents

v

Preface

Mathematics and its practitioners have come a long way since the days of drawing polygons in the sand with a stick. Although this cannot be said of all our degree courses, there is nevertheless an increasing realisation in higher mathematics education that current computing technology can open new doors for students and tutors alike. This book arose out of a largely successful attempt to complement traditional mathematical courses with one which took this opportunity seriously.

First year students at a UK university are expected to acquire a wide range of mathematical skills—the ability to argue logically, absorb new concepts, calculate accurately, translate everyday problems into appropriate mathematical language, construct mathematical models and to assess the approximations made. We chose to use the popular and powerful computer package MATLAB® to help promote some of these skills. It provided a convenient way to help students understand things graphically, to see the wood rather than the trees in complex problems and to give access to more realistic modelling situations.

We chose MATLAB rather than one of the increasingly sophisticated and algebraically based packages because of the very gently sloped learning curve involved. MATLAB allows the student to graduate smoothly from the functionality of a hand calculator, through increasing use of powerful numerical and graphical facilities towards a high level programming capability. The latter point was considered a bonus in that it provided a possible access route to programming for students with no prior computer background. At the very least, students with no keyboard skills at all can acquire a degree of familiarity with an essential modern tool, the computer.

The course, and this book, were designed for students coming to grips with a typical first year honours mathematics course at a UK university.

xi

In our case, students had already completed the first semester of core units and so already had a basic knowledge of calculus, complex numbers, vectors and matrices. In the book, we assume that the reader has a reasonable level of skill in calculus but only limited familiarity with the other topics. The typical student will be in the process of extending this base to include some selection of topics such as elementary statistics, mechanics, linear algebra, number theory, differential equations, Fourier series and so on. The book is thus intended to help motivate new topics and to build on old ones.

Like Gaul, the book is divided into three parts. Part one comprises a very elementary 'hands-on' introduction to the features of MATLAB followed by a series of methods chapters. In these the reader is taken through a range of mathematical ideas and given 'on the job' training in those MATLAB techniques which are expected to be of particular value in the ensuing project chapters. Thus all the standard programming structures and MATLAB commands are introduced through work on: matrices; whole numbers and elementary number theory; graphing plane curves; data fitting and approximations to functions using least squares techniques; simulation of random distributions; and ordinary differential equations. In this way, the student learning how to use MATLAB is taken through mathematics which is (or should be!) interesting for its own sake.

Part two contains a variety of projects, termed 'Investigations', which build on the earlier ideas. Matrices are applied in the context of magic squares, permutations and the solution of linear systems; manipulation of whole numbers is applied to greatest common divisors of random sets of numbers, primality testing and card shuffling; approximation techniques are applied to solution of nonlinear equations and interpolations; and so on. In each case, an exploratory attitude is encouraged, backed up with plenty of explicit exercises, both purely computational and more mathematical in nature.

Finally, Part three contains a number of 'Modelling Projects' in which the reader is invited to employ some of the skills developed in Part one. By its nature, mathematical modelling is a rather open-ended process and requires a certain degree of mathematical maturity that a first year student may not yet have attained. Nevertheless, we feel that the availability of techniques to which MATLAB gives access, and the very great importance of modelling as an applied mathematical skill, mean that this is an opportunity not to be missed. In practice, we have found that students cope well with these challenges.

At Liverpool, we required students to work through the preliminary material (Part one, taking six weeks), and allowed them to choose a total of three projects from Parts two and three, with at least one from each part. Two weeks seemed to be a good time to allow for the completion of one project, so that the whole course was twelve weeks long.

We have of course striven for uniformity in important matters throughout the book. But a discerning reader will detect three different styles in the project work of Parts two and three, providing a measure of variety which we feel is entirely healthy.

We have provided appendices which list MATLAB commands, give some information on symbolic calculations (not used explicitly in the material of the book) and MATLAB resources, and list the available M-files chapter by chapter.

Using the book

The book will prove useful in a number of contexts. Firstly it can be used, as it stands, to deliver a complete course unit. Secondly, the book should prove useful to course designers with slightly differing requirements. In this case the various examples of project work will provide a convenient source of material and stimulate the creation of further material tailored to the local need. Thirdly, the book can serve as a self-contained tutor for the enthusiastic individual who is not following any formal course structure.

In every case, the reader is intended to work through the book while sitting at the computer keyboard, although there are also mathematical exercises to be done off-line. A preliminary skim through this preface and Chapter 1 will help orient the newcomer before plunging in.

Readers who already have some experience with MATLAB might well wish to jump straight into Chapter 2. If in doubt, readers can quickly brush up their skills with the exercises at the end of Chapter 1.

Copies of all the 'M-files' to which the text refers are freely available. Details of how to obtain these are given in the Appendix. Partial solutions and hints are available to course tutors in electronic form on request from the publisher.

MATLAB is available on a wide variety of platforms. For definiteness, the book assumes the reader has access to MATLAB within a Microsoft Windows environment. Should this not be the case, readers may experience some small inconvenience in the early stages while adapting file-handling and editing instructions to suit their own installation, but the M-files should all run correctly and the material of the book itself is platform independent. Course providers might wish to make available a brief summary of key points where the local reader might otherwise go astray.

Acknowledgments

We are very grateful to colleagues at Liverpool who have helped to set up this course and have provided input to the material. These are Nigel Backhouse, Eric Edmond, Toby Hall, Neil Kirk, Dick Wait, Neville Waters. We are especially grateful to the students who have taken the course over several years; their efforts have detected possible ambiguities in the project work and helped to make the material more user-friendly — and, we hope, correct — than it was at the beginning. Any remaining faults are of course our responsibility. PG is also grateful to Brown University in Providence for its generous hospitality during part of the writing period of this book, and to the Fulbright Commission for a travel grant.

Liverpool, October 1998

Ke Chen
Peter Giblin
Alan Irving

Part one
Foundations

1

Introduction

1.1 First steps with MATLAB

If you haven't already done so, you should start MATLAB by *double-clicking* the relevant *icon* with your *mouse* or by asking a friend sitting next to you to show you how. Asking a friend is often the quickest way to obtain help and, in what follows, we will encourage you to take this route when all else fails. If 'clicking' and 'icons' mean nothing to you, you may need some extra help in getting started with Windows. It might also be that your system doesn't use *Microsoft Windows* and that simply typing `matlab` will do the trick. For example, if you are using a *Unix* system of some kind this may be the case.† If all goes well you will see a MATLAB prompt

```
>>
```

inviting you to initiate a calculation. In what follows, any line beginning with >> indicates `typed input` to MATLAB. You are expected to type what follows but not the >> prompt itself. MATLAB supplies that automatically.

1.1.1 Arithmetic with MATLAB

MATLAB understands the basic arithmetic operations: add is +, subtract is -, multiply is * and divide is /. Powers are indicated with ^, thus typing

```
>> 5*5+12^2
```

results in

---

† We will make some further remarks, where appropriate, about editing and file handling in a non-Windows environment.

3

```
ans =
```

```
169
```

If you typed the above line and nothing happened, perhaps you omitted to press the <Enter> key at the end of the line. The laws of precedence are built in but, if in doubt, you should put in the brackets. For example

```
>> 8*(1/(5-3)-1/(5+3))
```

```
ans =
```

```
3
```

The sort of elementary functions familiar on hand calculators are also available. Try

```
>> sqrt(5^2+12^2)
```

and

```
>> exp(log(1.7))
```

What do think `sin(pi/2)` will produce? Try it.

In fact MATLAB has the value of $\pi = 3.1415926\ldots$ built in. Simply type in `pi` whenever you need it. Try the following:

```
>> pi
>> format long
>> pi
>> format short
```

MATLAB retains considerably more significant figures of accuracy than suggested by the default setting which is given by `format short`.

1.1.2 Using variables

You can assign numerical values to '*variables*' for use in subsequent calculations. Typing

```
>> x=3
```

produces

```
x =

    3
```

or you might want something more useful like

```
>> rad=2; ht=3;
>> vol=pi*ht*rad^2

vol =

    75.3982
```

Note that the first line had two 'commands' on it, neither of which seemed to produce a result! When MATLAB encounters an instruction followed by a semi-colon ; it suppresses any visual confirmation. It really does obey the instruction but keeps quiet, as you can check with

```
>> rad=4;
>> rad

rad =

    4
```

This is useful if you want to avoid cluttering up the screen with intermediate results. Watch out for semi-colons in what follows.

Remember that each variable must somehow be assigned a value before you can make use of it in further calculations. For example if you have followed the above examples and now type

```
>> f = x^2 + 2*x*y + y^2
```

you should get a result something like

```
??? Undefined function or variable y
```

This is self-explanatory. If you now type y=4; and then repeat the calculation of f you should have more success.

Incidentally, a quick way of repeating a previous MATLAB instruction is to press the 'up-arrow' key (↑) until you recover the command you want. Try it now. If the original instruction was not quite correct, or if you want to develop a new instruction from a complex but similar previous one, you can use the same trick. Recover the command which you want then use the sideways arrows ($\leftarrow$ and $\rightarrow$) together with the delete key to edit the old command suitably. As an exercise, try using your previous work to calculate the volume of a right-circular cylinder with radius 2 and height 1/4. Did you get π?

If you have difficulty remembering the names of variables which you have assigned, you can try typing who or whos. Try them both now. Do you recognise the variables listed?

1.2 Vectors and plots

One of the pleasures in learning to use MATLAB is discovering the simplicity of plotting things. The basic principle is:

(i) select a sequence of x-values that is, a vector of values;
(ii) evaluate $y = f(x)$, that is, obtain a corresponding vector of y-values
(iii) plot y vs x.

Before doing this, it is worth spending a moment learning something about how MATLAB deals with vectors.

1.2.1 *Vectors*

Type in the following examples which all result in vector-valued variables. Pause to think about the result in each case.

```
>> u=[2,2,3]
>> u=[2 2 3]
>> v=[1,0,-1]
>> w=u-2*v
>> range=1:13
>> odd=1:2:13
>> down=20:-0.5:0
>> even=odd+1
>> xgrid=0:.05:1; x=xgrid*pi
>> y=sin(x)
```

The first two lines demonstrate that elements of a vector can be separated by spaces or commas. If you are worried about inserting blank spaces by accident you can stick to the comma notation. Thus [1+1 2 3] is the same as [2,2,3], whereas [1 +1 2 3] is the same as [1,1,2,3]!

Note that vectors can be of any length. They can be row vectors as here, or column vectors like

```
>> w'
ans =

     0
     2
     5
```

where the apostrophe denotes transpose (T). In MATLAB, vectors are treated simply as a special case of matrices which you will learn much more about in the next chapter.

Notice what happens when the displayed vector is too long to fit on a line. MATLAB just displays as many elements as it can and then puts the rest on the following lines. The elements in a row vector are treated as 'columns'.

An elementary function of a vector x, such as `sin(x)`, is also a vector of the same kind. We can use this fact to create plots of functions as shown in the next section.

MATLAB knows how to multiply matrices of compatible size. This will be discussed in greater detail in the next chapter. For now, try typing

```
>> w*w'
>> w'*w
>> u*w'
>> u*u
```

Can you make sense of the results? Why did the last one not work?

Now suppose you want a set of values z given by $z = y^2$, where y is the vector of values already assigned. From the above experiments you will realise that

```
>> z=y*y
```

is not understood by MATLAB.

```
>> z=y*y'
```

is understood but is evaluated as the scalar product $\mathbf{y} \cdot \mathbf{y}$! What you need to do, to force MATLAB to multiply things *element-by-element*, is to type

```
>> z=y.*y
```

where the . inserted before the ∗ symbol is the key feature forcing element-by-element operation. Similarly, u./v and y.^2 are understood as element-by-element operations with vectors of the same size.

1.2.2 Plotting things

Type whos at this point to verify that you have vectors x and y defined as above. They should both be 1×21 matrices (that is, row vectors).

Plotting is easy. Just type

```
>> plot(x,y)
```

and sit back. A nice simple graph of $y = \sin x$ vs x will magically appear. The axes are chosen automatically to suit the range of variables used. This is the simplest possible case. In later work you will want to do more elaborate things. For now, try the following:

```
>> title('Graph of y=sin(x)')
>> xlabel('x')
>> ylabel('y')
>> y1=2*x;
>> hold on
>> plot(x,y1,'r')
```

You can probably figure out the significance of each of these commands. For example, y1=2*x defined new function values $y = 2x$, hold on told MATLAB to keep the same graph and plot(x,y1,'r') plotted the next graph on top. Note that the axes were adjusted† and the second curve was plotted in red.

In the above examples of plot, MATLAB joined up the 21 points with straight-line segments. Should you not want this, you can specify plotting points using a choice of symbols as follows. Try this

```
>> hold off
>> plot(x,y,'+')
>> plot(x,y,'g*')
>> plot(x,y,'w.')
```

---

† Assuming you are using MATLAB version 4. There are a number of small differences between this and earlier versions, particularly with graphics commands.

Did MATLAB do what you expected? Did you remember to use the ↑ key to reuse previous commands? If you need some help with how to use any MATLAB instruction you can type for example

```
>> help plot
>> help hold
>> help sin
```

and so on.

1.3 Creating and editing script files

Once you get going, you may find it tiresome to keep reentering the same, or similar, sequences of commands. Fortunately, there is a simple way round this: you simply store any frequently used sequence of commands in a file called a '*script*' or '*M-file*'. You can then invoke this list of commands as often as needed.

For example, in a particular session you might want to find the distance between two points A and B whose position vectors are given by $\mathbf{a} = (1, 0, -2)$ and $\mathbf{b} = (2, 3, 1)$ respectively. Knowing that the vector displacement between them is

$$\mathbf{d} = \mathbf{b} - \mathbf{a}$$

and that

$$|d|^2 = \mathbf{d} \cdot \mathbf{d},$$

you might use the following sequence of MATLAB instructions:

```
>> a=[1,0,-2];
>> b=[2,3,1];
>> d=b-a;
>> dd=d*d';
>> dist=sqrt(dd)
```

to solve that particular problem. This is fine, but suppose you have a set of five points and want to check which pair is the closest together? You would obviously want to store as many as possible of these steps in a 'script' (file) to reuse as required.

1.3.1 Editing and saving a text file

We first need to review file-handling and editing.

Non-Windows users

If you are *not* using Microsoft Windows, you will at this point need to make some slight alteration to procedures. However, no matter how MATLAB has been installed on your system, there will undoubtedly be a *text editor* of some sort. Assuming it is called `edit`, the easiest way to invoke it from MATLAB is probably to type

```
>> !edit fname
```

where fname is the name of a text file which either exists or will exist, by the time you have finished! If that doesn't work, consult someone knowledgeable about your system setup or ask a patient friend.

Windows users

Windows comes with its own basic text file editor called *Notepad*, whose icon is usually found within the Accessories Group. A typical MATLAB setup within Windows makes direct use of this accessory so we will concentrate on this method. To create and edit a new file called `myfile.m`, *from within MATLAB*, proceed as follows:

(i) In the `MATLAB Command Window` menu, click on File.
(ii) Click New then M-file.
(iii) Within Notepad which you have now started, you can type some lines, for example

```
% myfile.m
% It doesn't do very much, just identifies itself.
% These 3 lines are comment lines which MATLAB ignores.
disp(' I am an M-file')
```

(iv) Click File then Save As.
 (v) Within the box File Name which is open waiting, type in `myfile.m`.
(vi) Click OK.

You have now created a file which MATLAB can find and use. You hope!

Back in the `MATLAB Command Window` you can now ask MATLAB whether it can find the file. Type

```
>> type myfile
```

and you should see a list of the lines which you typed in. If not, go back to step (i) and start Notepad again by clicking File in the `MATLAB Command Window` followed this time by Open M-file. You should see an entry corresponding to `myfile.m`. If not, you should go right back to the beginning of this section, perhaps with someone else watching your steps this time.

1.3.2 Script files

If all is well, you now have your first example of a script file. To use it you simply type

```
>> myfile
```

whereupon you should see something like

```
 I am an M-file
```

Now for something more useful. We will set up an M-file to repeat the earlier instructions for getting the distance between two points. Proceed as above to create a new M-file called `distab.m` containing a few comment lines, such as

```
% distab.m
% Calculates the distance between two vectors a and b
% ...
```

including any more comment lines (those starting with %) which you may need to remind yourself how it works. Then the business lines

```
d=b-a;
dd=d*d';
dist=sqrt(dd)
```

Remember to save the file by clicking File then Save As as you did when creating `mfile.m`. Review that example if necessary. When you have finished editing the file, you might as well close Notepad by clicking File then Exit. You can leave Notepad open if you want, but it may get confusing if you leave too many Notepad windows open at once. When exiting, Notepad always reminds you if there are recent changes to be saved.

Now pick values for the vectors **a** and **b**, if you haven't already done so

```
>> a=[1,0,-2];
>> b=[2,3,1];
```

Then find the distance by simply typing

```
>> distab
```

Did it work? If not, go back into Notepad and keep trying.

Try it for a different choice of points A and B, perhaps for a pair whose distance you can trivially check, for example $A \equiv (1,2,3)$ and $B \equiv (1,1,3)$.

```
>> a=[1,2,3];
>> b=[1,1,3];
>> distab
```

To find out what M-files you have created, or which others MATLAB already contains, use the command `what`. If you now want to check the purpose of an M-file you can use `help` (just as with any MATLAB instruction)

```
>> help myfile
>> help sqrt
>> help sin
```

The command `help` displays the initial *comment* lines at the top of an M-file. This is why it is always *good practice* to include comment lines (starting with %) at the top of an M-file. It is also a good idea to include the name or title in the first comment line.

Note that `myfile` is the name of the MATLAB instruction (what you type to use it), whereas `myfile.m` is the *name of the file* containing its definition.

1.3.3 Function files

It was tedious to have to assign the two vectors each time before using the above '*script*'. You can combine the assignment of input values with the actual instruction which invokes the M-file by using a *function M-file*. Not only that, but you can at the same time assign the answer to a new variable, that is, you can design a function M-file `distfn` such that typing

```
>> dab=distfn([1,2,3],[1,1,3]);
```

or

```
>> a=[1,2,3]; b=[1,1,3]);
>> dab=distfn(a,b);
```

assigns the correct distance to the variable `ab` without any further ado.

Here is how to modify the script M-file `distab.m` to become the function M-file `distfn.m`. We will assume that you didn't leave Notepad open and so will have to start editing from scratch.

(i) In the `MATLAB Command Window` menu, click on <u>F</u>ile.

(ii) Click <u>O</u>pen `M-file` to start Notepad and look for the list of available M-files, that is, ones ending with `.m`. (If necessary, change the `*.txt` ending showing in the File <u>N</u>ame box to `*.m`).

(iii) Select `distab.m`.

(iv) Now you can make changes. First change the comment lines to reflect the new name and purpose and then the modify the MATLAB instructions so that your file looks like this:

```
% distfn.m
% Calculates the distance between two vectors a and b
% Usage:-
% dist=distfn(a,b)
% input:  a,b (position vectors)
% output: distfn is the distance between them
function dist=distfn(a,b)
d=b-a;
dd=d*d';
dist=sqrt(dd);
```

(v) Click `Save` <u>A</u>s.

(vi) Within the box `File` <u>N</u>ame which is opened waiting, type in `distfn.m`.

(vii) Click `OK`.

Back in the MATLAB Command Window, you can now type `help distfn` to remind yourself how to use it. You type

```
>> dist=distfn([1,1,1],[2,2,2])
```

or

```
>> dist=distfn(a,b)
```

to assign the required distance to the variable `dist`. If you got more than one number flashing up, perhaps you forgot some of the semi-colons with the function. If so have another look at it using Notepad.

Function M-files (or 'M-functions'), that is, M-files whose first non-comment line starts `function` ..., have a very important feature. Aside from the name itself (`distfn` in this case), all the other variables (`a`, `b`, `dd` etc.) are purely internal to the function. This can help reduce confusion with other calculations and variables which you may have used. Check this by typing

```
>> who
>> clear
>> who
>> dist=distfn([1,1,1],[2,2,2])
>> who
```

The command `clear` clears out all previously defined variables. After executing the function with `dist=distfn(..)`, no trace of the internal variables remains in your MATLAB session.

1.3.4 Diary files and saving things

You will sometimes find it handy to keep a copy of what you produce on the screen. You might want to print out bits of it later (see the next section). This is done very easily. To see how it works, type the following

```
>> diary sect1.txt
>> % Beginning of section 1
>> % now some commands
>> myfile
>> dist=distfn([1,1,1],[2,2,2])
>> diary sect2.txt
>> % I want this stored somewhere else
>> x=0:.1:1;
>> y=x.*x
>> plot(x,y)
>> diary off
```

The session will proceed fairly normally. The command `diary fname` tells MATLAB to put a copy of the text output (numbers and letters) in the file `fname`. Typing `diary off` switches it off. The above example puts some output into one file and some into another. You can now use Notepad to have a look at what was generated and, if needed, edit it. Invoke Notepad as usual by clicking <u>F</u>ile and try to <u>O</u>pen files with the ending `*.txt`. You should see `sect1.txt` and `sect2.txt` in the list.

When you look in `sect2.txt` you may be disappointed to find no plot of x^2 there! This is because plots and other graphical images can't easily be represented as text files. The next section shows how to get round this.

When you end a MATLAB session all current variables and their values are lost. Usually this doesn't matter. The effect of stopping and restarting a session is the same as typing

```
>> clear
```

that is, you start with a clean sheet. Should you actually wish to save what you are doing for another time you can type, for example,

```
>> save monday
```

or

```
>> save monday x,y,mymatrix
```

where the second version saves only the explicitly mentioned variables in the specified file **monday**. You can then reload things the next day or whenever with

```
>> load monday
```

Remember that only the variables themselves that is, their current values, are saved using these techniques. Any formulae which you have typed in will be lost unless you have entered and saved them in an M-file.

1.4 Getting hardcopy of things

1.4.1 Windows users

An advantage of using a Windows environment is that printing text files, such as diary files and M-files, is performed in a completely standard way in all applications. MATLAB is normally set up to take full advantage of this. The same is true for the plot images which MATLAB produces.

To get a copy of an M-file or other text file just open it up using Notepad in the usual way (see previous section). Click on File and then Print. That's it! If it isn't, then perhaps you need to check Print Setup on the same menu to see where the output is being sent. If necessary, you may need to check with your local expert. Usually things have already been set up so all output goes to the most convenient local printer.

Table 1.1. *Summary of basic commands introduced so far.*

plot(x,y)	plot(x,y,'*')	plot(x,y,'+g')
title('The title')	xlabel('x label')	xlabel('x label')
sqrt(x)	sin(x)	exp(x)
hold on	hold off	
x=-1:.2:1	y=x.*x	dotprod=x*y' [a]
format long	format short	
help sqrt	help myfile	save fname
diary file1.txt	diary off	load fname

[a] (' ≡ transpose)

For plots, you go about getting a hardcopy in the same way but you use the <u>F</u>ile and <u>P</u>rint buttons on the Window containing the figure (usually labelled Figure No. 1).

1.4.2 Non-Windows users

If you don't have a Windows system or want to by-pass it, you can usually get away with something like

```
>> !print fname.txt
```

or use the appropriate print command of the underlying operating system (for example `lp` or `lpr` for Unix).

For plots, the MATLAB command `print` is usually set up to print a copy of the *current* figure on the default printer. If this is not so, you may make some progress by following the advice in `help print`. Failing that, try asking your patient friend.

Exercises

Before proceeding, check your level of skill by completing all of the following short exercises. If you can't remember some of the commands have a look at the list in Table 1.1. If necessary, go back and reread the relevant section.

1.1   Find the sum of the first four terms in the sequence

$$\frac{1}{2\times 3}, \quad \frac{2}{3\times 4}, \quad \frac{3}{4\times 5}, \cdots$$

1.2   Define a vector t with values evenly spaced by 0.2 between 0
       and 6 inclusive. Now use this to obtain plots of

$$f(t) = \sin(\pi t)$$

and

$$g(t) = \exp(-t)\sin(\pi t)$$

on the same graph with the first in green and the second in
yellow. If you are unsure of the MATLAB functions needed,
use `help exp` etc. Enhance the graph by adding a white line
corresponding to $y = 0$.

1.3   Use the editor to create an M-file which deduces the length of
       each side of a triangle ABC whose vertices have position vectors
       $\mathbf{a} = [1, 2, 3]$, $\mathbf{b} = [2, 3, 4]$ and $\mathbf{c} = [3, 4, 5]$.

2
Matrices and Complex Numbers

2.1 Vectors and matrices

2.1.1 Vectors

We recall briefly how to enter vectors. Let

$$\mathbf{a} = (-1, 2, 4) \quad \text{and} \quad \mathbf{b} = (1.5, 2, -1).$$

To assign these vectors to variables $\mathbf{a}$ and $\mathbf{b}$ type either

```
>>   a = [-1 2 4]
>>   b = [1.5 2 -1]
```

or

```
>>   a = [-1,2,4]
>>   b = [1.5,2,-1]
```

Thus spaces or commas can be used.

One way of finding the *dot* or *scalar product*, of two vectors, say $\mathbf{a} \cdot \mathbf{b}$, was introduced in §1.2. Here is another which uses the important idea of *element by element multiplication* also introduced in Chapter 1. Typing

```
>>   c=a.*b
```

where there is a *dot* before the multiplication sign *, multiplies the vectors $\mathbf{a}$ and $\mathbf{b}$ element-by-element: in this case, $\mathbf{c} = (-1.5, 4, -4)$. The dot product is then obtained by summing the elements of $\mathbf{c}$:

```
>>   sum(c)
```

gives $\mathbf{a} \cdot \mathbf{b} = -1.5$. Similarly

```
>>   sqrt(sum(a.*a))
```

18

gives the magnitude of **a**. In fact, the MATLAB command, `norm`, will directly find the magnitude of a vector. To find the angle θ between **a** and **b**, we can use $\theta = \cos^{-1} \mathbf{a} \cdot \mathbf{b}/|\mathbf{a}| \, |\mathbf{b}|$. In MATLAB, $\cos^{-1}$ is `acos` so the the complete calculation, performed this way, would be written

```
>>  theta = acos( sum(a.*b) / sqrt(sum(a.*a)*sum(b.*b)) )
```

and gives $\theta = 1.693$ radians approximately.

2.1.2 Matrices

The matrix

$$A = \begin{pmatrix} -1 & 1 & 2 \\ 3 & -1 & 1 \\ -1 & 3 & 4 \end{pmatrix}$$

is entered in MATLAB as

```
>>  A=[-1 1 2;3 -1 1;-1 3 4]
```

with semicolons separating the row vectors, or as

```
>>  A = [-1  1  2
    3  -1  1
    -1  3  4]
```

with different rows separated by <Enter>, therefore appearing on different lines. The MATLAB prompt >> will not reappear until you have finished defining the matrix by closing the bracket]. If you accidentally make one row have more entries in it than another row then you will get an error message. Note that, if you wish, you can insert commas between the entries which lie in the same row:

```
>>  A=[-1,1,2;3,-1,1;-1,3,4]
```

If you find the line spacing in MATLAB output is over-generous for your taste you can suppress extra spacing with the command

```
>>  format compact
```

At some stage, have a look at `help format` to see what else you can do to control the look of MATLAB output.

The equations

$$\begin{aligned} -x_1 &+& x_2 &+& 2x_3 &=& 10 \\ 3x_1 &-& x_2 &+& x_3 &=& -20 \\ -x_1 &+& 3x_2 &+& 4x_3 &=& 40 \end{aligned} \qquad (2.1)$$

can be written as

$$Ax = b \qquad (2.2)$$

say, where $\mathbf{x}$ is the column vector $(x_1, x_2, x_3)^\top$ and $\mathbf{b}$ is the column vector $(10, -20, 40)^\top$ of right-hand sides of the equations. In MATLAB, we can write $\mathbf{b}$ as the transpose of a row vector by using

```
>> b=[10  -20  40]'
```

In order to solve linear equations $Ax = \mathbf{b}$, where the determinant of A is nonzero, we can use the *inverse* A^{-1} of A, that is, the matrix such that $AA^{-1} = A^{-1}A = I$, the 3×3 identity matrix. To discover whether the determinant is zero, type

```
>> det(A)
```

which gives the answer 10. The solution for $\mathbf{x}$ is then obtained by

$$\mathbf{x} = A^{-1}Ax = A^{-1}\mathbf{b},$$

which in MATLAB is

```
>> x=inv(A)*b
```

displaying the result as a column vector $\mathbf{x}$ (in this case $(1, 19, -4)^\top$).

Thus matrix products are obtained by using the multiplication symbol * but the matrices must be of compatible sizes or MATLAB produces an error. For example b*A is meaningless since the number of columns of $\mathbf{b}$, namely 1, does not equal the number of rows of A, namely 3. You can test matrix multiplication by

```
>> C=A*A
>> det(C)
>> D=A^3
>> det(D)
```

Thus $C = A^2$, $D = A^3$ and, by the well-known rule for square matrices of the same size $\det(P)\det(Q) = \det(PQ)$, the determinants are 100, 1000 respectively.

Actually, one does not usually solve linear equations such as equations (2.1) or (2.2) by first finding the full inverse of the corresponding matrix. One applies linear operations to the augmented matrix formed from A and $\mathbf{b}$. MATLAB can also perform the calculation in this way.† You simply type

† See Chapter 16 for a more detailed discussion of the solution of linear systems.

```
>>   x = A\b
```

Again, the matrices must have compatible dimensions. Try it and check your answer by forming the product $A\mathbf{x}$ or by finding the 'residual' $\mathbf{r} = \mathbf{b} - A\mathbf{x}$.

You can also add two matrices *of the same size*. MATLAB doesn't even complain if you add a scalar to a matrix. In fact

```
>>   E=A^2+2*A+1
```

produces the matrix

$$E = A^2 + 2A + \begin{pmatrix} 1 & 1 & 1 \\ 1 & 1 & 1 \\ 1 & 1 & 1 \end{pmatrix}.$$

So don't assume that 1 means the identity matrix.† The identity matrix and the matrix all of whose elements are 1 can be called up by typing

```
>>   diag([1 1 1])
>>   ones(3,3)
```

so that **diag** makes a diagonal matrix with the given diagonal entries and **ones** makes a matrix of 1s of the given size. What does

```
>>   diag(ones(1,3))
```

give? Note the brackets when using the function **diag**. The identity matrix is actually built into MATLAB as (e.g. for 3×3)

```
>>   eye(3)
```

As already noted, one may wish to solve the equations $A\mathbf{x} = \mathbf{b}$, by forming the corresponding *augmented matrix*

```
>>   F = [A b]
```

In fact, this method is applicable even when the determinant is zero or when the matrix A of the equations is not square. The procedure is to reduce the augmented matrix to 'row reduced echelon form', that is, to use row operations to produce a matrix in which there is a leading 1 in each row, the entries before, above and below which are all zeros. We shall not use this in any important way here, so if you have not met the idea just skip over the remainder of this section. The row reduced echelon form is used in the project of Chapter 8 on magic squares.

---

† Unfortunately, typing **A*1** produces A again, suggesting that in multiplication MATLAB thinks 1 means the identity matrix!

Typing

```
>>  G=rref(F)
```

produces the row reduced echelon form of F. In this case it is

$$G = \begin{pmatrix} 1 & 0 & 0 & 1 \\ 0 & 1 & 0 & 19 \\ 0 & 0 & 1 & -4 \end{pmatrix}.$$

This means that the original equations are *equivalent* to the equations whose augmented matrix is the matrix G, and these equations are simply $x = 1, y = 19, z = -4$ so the solutions can be read off.

Don't forget the round brackets when using a MATLAB function such as `rref`. Thus if you type in a matrix directly you must still include them, as in

```
>>  rref([1 2 3; 4 5 6; 7 8 9])
```

Note that this produces a single row of zeros (the last row), indicating that the rows of the original matrix are linearly dependent. You can check this by finding the determinant, which is zero.

You may also care to try typing

```
>>  rrefmovie(F)
```

which gives an animated demonstration of the row reduction process.

2.1.3 Eigenvalues and eigenvectors

If A is an $n \times n$ matrix, λ is a number and $\mathbf{x}$ is a nonzero $n \times 1$ (column) vector, such that

$$A\mathbf{x} = \lambda\mathbf{x},$$

then $\mathbf{x}$ is called an *eigenvector* of A and λ is the corresponding *eigenvalue*.† The eigenvalues can also be thought of as the numbers λ such that $\det(A - \lambda I) = 0$, where I is the $n \times n$ identity matrix.

For example, try typing

```
>>  A=diag([1 2 3])
>>  eig(A)
>>  P=[1 2 3;4 5 6; 5 7 8]
>>  det(P)
```

---

† Geometrically, this means that the linear map associated with A sends $\mathbf{x}$ to another vector along the same line through the origin as $\mathbf{x}$.

```
>>   B=inv(P)*A*P
>>   eig(B)
>>   [Y,D]=eig(B)
```

The eigenvalues of A should be simply the diagonal entries 1, 2 and 3. Since the matrix P is nonsingular $(\det(P) = 3)$, the matrices A and $B = P^{-1}AP$ are 'similar' and so have the same eigenvalues. Note that MATLAB may order them differently. The last line finds the eigenvectors of B as well: the matrix Y has columns which are eigenvectors and the diagonal matrix D has *diagonal* entries which are eigenvalues. The eigenvector corresponding to the first diagonal element of D is the first column of Y, and so on. In this case, all eigenvalues and eigenvectors are real. For an example where this does not happen, try the matrix A of §2.1.2.

2.2 Complex numbers

MATLAB has the symbol i $(\sqrt{-1})$ built in together with the rules of complex arithmetic. MATLAB also reserves the symbol j for $\sqrt{-1}$. Try typing

```
>>   clear
>>   i
>>   j
>>   i*i
```

and see what happens.

Warning: when dealing with complex numbers, there is a risk that MATLAB will misunderstand the symbol i for the square root of –1. This can happen if i has been used recently for an indexing variable, usually an integer (we shall meet indexing in `for` loops later; see §3.1). To safeguard against this, just type

```
>>   clear i,j
```

before starting work on complex numbers, and avoid using i for anything else!

You will already have seen some complex numbers if you found the eigenvalues and eigenvectors in §2.1. MATLAB can handle complex numbers easily. For example,

```
>>   a=1+i;  b=2-3i;
>>   c=a*b
```

```
>>   d=sqrt(a)
```

produces the answers $c = 5 - i, d = 1.0987 + 0.4551i$. Note that just one
square root is shown. Likewise

```
>>   (-1)^(1/2)
>>   (-2+2i)^(1/3)
```

produce respectively the answers $i, 1 + i$. The other square or cube
roots must be obtained by multiplying by –1 or by cube roots of unity,
respectively.

The modulus, argument ($\geq -\pi$ and $\leq \pi$ radians) and real parts of an
already specified complex number a are obtained by

```
>>   abs(a)
>>   angle(a)
>>   real(a)
```

respectively. The command

```
>>   imag(a)
```

produces the imaginary part without the i attached. For example with
$a = 1 + i$ the answer is 1.

```
>>   conj(a)
```

produces the complex conjugate $\bar{a}$ of a. (If $a = x + iy$ where x and y are
real, then $\bar{a} = x - iy$.) Thus

```
>>   a*conj(a)-abs(a)^2
```

should always give answer 0 (but might give a very small answer such
as $1.7764\text{e-}015$, that is, 1.7764×10^{-15}).

Try also

```
>>   exp(i*pi)
```

which produces $e^{i\pi} = -1$.

2.3 Population dynamics: the Leslie matrix

There are many uses of matrices and their eigenvalue properties in eco-
nomics, life-sciences and probability theory as well as in the physical
sciences. A very simple example is in the following discrete time model

for the age structure of the population of a country or other large community. The basic idea is to take a vector representing the age distribution in one year, construct a matrix of transition probabilities from one year to the next and then, by matrix multiplication, predict the probable age distribution for the next year. Predictions for subsequent years are obtained by further matrix multiplication. The hard bit is to make a model for the probability matrix! Here is an example.

In some given year, we count the number of people in age-bands 0–5, 6–19, 20–59 and 60–69. Further subdivisions are of course possible with a little more work. We then make the following, fairly drastic, simplifying assumptions

- *Within* one age-band, the age distribution is constant, that is, there are the same number of people in each year group.
- We do not consider anyone who lives beyond 69!
- Deaths only occur in the 60–69 age-band at the rate of $d_A\%$ per annum and in the 0–5 band at $d_I\%$ per annum.
- Births are due to people in the 20–59 age-band at the rate of $b\%$ per annum.

With these assumptions, we want to relate the column vector

$$\mathbf{N}(2) = (n_1(2), n_2(2), n_3(2), n_4(2))^\top$$

of populations in the four age-bands in year 2 to the column vector

$$\mathbf{N}(1) = (n_1(1), n_2(1), n_3(1), n_4(1))^\top$$

of populations in year 1. Recall that, as above, $\top$ stands for transpose of a vector or matrix.

To see how to obtain this relationship, consider Table 2.1.

The various columns contain: the age range within the band; the symbol n_i for the current number of persons in that band; the number of year groups (for example, ages 0, 1, 2, 3, 4, 5 make six year groups); the number that, according to our rules, die in that year; the consequent number (the rest) who survive the year; the number of these survivors who leave the band (graduating to the next one or, in the case of the age-band 60–69, passing out of our calculations); and the number who enter the band, either through being born into it or through graduating from the one below.

Of course, each band's leavers become the next band's entrants. According to the simplifying assumptions, the number in each individual

Table 2.1. *Construction of a Leslie matrix.*

Age-band	No. in band	Year grps.	Die	Survive	Leave band	Enter band
0–5	n_1	6	$\frac{d_I}{100}n_1$	$\left(1-\frac{d_I}{100}\right)n_1$	$\left(1-\frac{d_I}{100}\right)\frac{n_1}{6}$	$\frac{b}{100}n_3$
6–19	n_2	14	0	n_2	$\frac{n_2}{14}$	$\left(1-\frac{d_I}{100}\right)\frac{n_1}{6}$
20–59	n_3	40	0	n_3	$\frac{n_3}{40}$	$\frac{n_2}{14}$
60–69	n_4	10	$\frac{d_A}{100}n_4$	$\left(1-\frac{d_A}{100}\right)n_4$	$\left(1-\frac{d_A}{100}\right)\frac{n_4}{10}$	$\frac{n_3}{40}$

year within an age-band is the same, so the 'leavers' are just the 'survivors' divided by the number of years covered by the age-band. In practice much narrower age-bands might be taken, to make this assumption more plausible.

It is now easy to see that the vector $\mathbf{N}(2)$ of populations of the various age-bands in year 2 is related to the corresponding vector $\mathbf{N}(1)$ for year 1 by a matrix equation

$$\mathbf{N}(2) = L\mathbf{N}(1),$$

where L is the so-called *Leslie matrix*

$$L = \begin{pmatrix} \frac{5}{6}\left(1-\frac{d_I}{100}\right) & 0 & \frac{b}{100} & 0 \\ \frac{1}{6}\left(1-\frac{d_I}{100}\right) & \frac{13}{14} & 0 & 0 \\ 0 & \frac{1}{14} & \frac{39}{40} & 0 \\ 0 & 0 & \frac{1}{40} & \frac{9}{10}\left(1-\frac{d_A}{100}\right) \end{pmatrix}.$$

Similarly,

$$\mathbf{N}(3) = L\mathbf{N}(2) = L^2\mathbf{N}(1),$$

and so on. In general, we have:

$$\mathbf{N}(t+1) = L\mathbf{N}(t).$$

Given data for $t = 0$, say, and some estimates of b, d_I and d_A, one can then predict the likely age structure of the population for a number of years ahead. This is very useful if you have to plan pension schemes, university sizes or day-care provision.

An M-file `leslie.m` is provided to allow you to explore these ideas. Type

```
>>   leslie
```

to obtain a 4×4 matrix L describing the evolution of the above model population. You are prompted to supply the birth rate (b) as a percentage per annum, for example 2.5, the infant mortality rate (for example $d_I = 1$) and the age-related death rate (for example $d_A = 10$) etc. Now set up an initial population, say

```
>>   N = [ 3.6 11.4 29.6 10.6]'
```

which is a very rough approximation to the 1996 UK population distribution in millions.

You should try the following:

Multiply **N** by L 4 times (`N1=L*N`, `N2=L*N1`, ...) to predict the UK population distribution in the year 2000 according to this model. What significant changes are there if any? Note that you can also type things like

```
>>   L^4
>>   N50=L^50*N
```

Type

```
>>   [Y D] = eig(L)
```

to obtain the eigenvectors and eigenvalues of the Leslie matrix L. Find the largest eigenvalue and its corresponding eigenvector **E**. You can compare the population vector after say 50 years, $L^{50}\mathbf{N}$, with this eigenvector **E** by

```
>>   N50=L^50*N
>>   N50./E
```

The second command divides the three elements of N50 by the three corresponding elements of **E**. The result should be three approximately equal numbers, showing that *the population after 50 years is roughly proportional to the eigenvector corresponding to the largest eigenvalue.*

What happens if the *initial* population distribution is proportional to this eigenvector?

Exercises

2.1   As a simple example of manipulation of complex numbers, consider the following, which will be covered in more general terms in the project of Chapter 13. (Note the warning given above, in §2.2; for safety you can type 'clear i' before starting.) Enter any complex number into MATLAB, for example

```
>>  z=3+4i
```

Now type

```
>>  z=(z^2-1)/(2*z)
```

Using the ↑ key, repeat the last command several times. Eventually the answers settle down—*converge*—to either i or $-i$; for the starting value $3 + 4i$ they settle down to i.

- Can you discover the rule which determines, from the starting z, whether the numbers will converge to i or $-i$? (You are not expected to prove the rule. That is covered in Chapter 13.)
- If you start with $z = 0$, MATLAB produces in turn $-\infty, 0, \infty, 0$, which doesn't make much sense since $z = \pm\infty$, if it gives anything at all, should give $z = \pm\infty$ again. But there are other starting values such as $z = 2$ which produce more reasonable nonconvergence. Which starting values for z produce nonconvergence? (Again only an experimental answer is expected.)

2.2   This exercise combines matrix work with editing files and using diaries. Make an M-file which contains the following (remember to leave spaces between the entries of a single row, or to insert commas there):

```
A=[1 1/2 1/3
1/2 1/3 1/4
1/3 1/4 1/5]
b=[1 0 0]'
det(A)
X=inv(A)*b
```

You could call this by some convenient reference name like ch2q2.m.

This M-file solves the equations with A as matrix of coefficients and the column vector **b** as the 'right-hand sides' of the equations. Note that the determinant of A is calculated first. Even if $\det A \neq 0$ the numerical solution may be very sensitive to small inaccuracies or changes in the parameters. You can

expect this if det A is small. This is just one example of an
ill-conditioned system.†

(a) Make a copy of `ch2q2.m`, say `ch2q2a.m` age-band edit the
file so that it solves the equations in which the elements 1/3 of
A are replaced by 0.333.
(b) Do the same replacing 1/3 by 0.33; you could call the M-file
`ch2q2b.m`.

In MATLAB, run your M-files and save the results in a diary.
Thus you type (in MATLAB) `diary ch2q2.txt` then `ch2q2` to
run this M-file, then `ch2q2a` to run the next M-file, and finally
`ch2q2b` to run the third one. When you have run them all type
`diary off`. Return to the editor and look at the file ch2q2.txt
to see what is in it. You should find that the results of solving
these similar sets of equations are very different. You could of
course add your own comments on this by typing them alongside
the results and print out the diary so that you have a 'hardcopy'
record.

2.3 Copy the M-file `leslie.m` and edit it as follows. The new M-file
is to be suitable for studying a population of cats described by
age bands 0–1, 2–5, 6–10 and 11–15 and obeying the following
modelling assumptions:

- No cats beyond age 15 are considered.
- Births only arise from the 2–5 band ($b\%$ per annum).
- Kitten deaths occur at the rate $d_K\%$ per annum in the 0–1
  band.
- All survive in the 2–5 band.
- Deaths from being run over occur in the 6–10 band at a rate
  $d_R\%$ per annum and from old age in the 11–15 band at $d_A\%$
  per annum.

An appropriate name for the resulting M-file is `cats.m`.

(a) Find the matrix generated when $b = 15$, $d_K = 5$, $d_R = 2$ and
$d_A = 30$. (Thus type, for example, `diary ch2q3.txt`, then run
the M-file by typing `cats` and entering the appropriate data,
then type `diary off`.)
(b) Assuming an initial population distribution $[20, 20, 20, 20]$,
does the overall population increase or decrease over a five-year

† See Chapter 16 for a discussion of ill-conditioned systems.

period? You can use `sum` to get the total of a vector. You can use `diary ch2q2q3.txt` again to include the result of finding the population during this period, and type in whether it is increasing or decreasing. Note that using the *same* diary name appends the new material to the old. It does not overwrite the original diary material. Don't forget to type `diary off` every time you have temporarily stopped writing to this diary.

(c) Find the largest (in magnitude) eigenvalue and corresponding eigenvector of the Leslie matrix C for the cat population evolution. Note that

```
>>  V=C(:,1)
```

picks out the first column $\mathbf{V}$ of a matrix C. This is very useful if you want to do something further with this column, such as compare it with another vector.

(d) Using the initial population in (b), show that the population vector ($\mathbf{N}$) after 50 years is approximately proportional to the eigenvector found in (c). Repeat for another initial population of your own choice. Recall that

```
>>  V1./V2
```

gives a vector containing the ratios of the elements of $\mathbf{V1}$ and $\mathbf{V2}$.

3

Whole Numbers

In this chapter, we shall introduce further MATLAB structures in the context of some properties of whole numbers.

3.1 A loop to calculate Fibonacci numbers

Type in succession

```
>> f = [1   1]
>> f(3) = f(1) + f(2)
>> f
>> f(4) = f(2) + f(3);
>> f
```

The last command produces the vector [1 1 2 3]. Thus f(1) refers to the first entry of the vector f, f(2) to the second, etc.

We can do this over and over in a loop:

```
>> f = [1   1];
for k = 1:15
f(k+2) = f(k+1) + f(k);
end
>> f
```

Note that MATLAB suppresses the >> prompt until the loop is completed by end. What the for...end loop does is to take in succession the values $1, 2, \ldots, 15$ for the variable k and to augment the vector f by a new entry $f(k+2)$ each time round. Thus $k = 1$ makes f emerge as [1 1 2]; $k = 2$ makes it emerge as [1 1 2 3], and so on. The semi-colon after the equation for $f(k+2)$ suppresses output during the 15 times the loop is executed.

31

Typing

```
>> plot(f,'*')
```

gives a plot of the values, placing an asterisk at each point $(i, \mathbf{f}(i))$. Just `plot(f)` joins these points up with straight segments, producing a steeply sloping curve. The entries of the vector $\mathbf{f}$ are the *Fibonacci numbers*, $1, 1, 2, 3, 5, 8, 13, 21, 34, \ldots$. The rule for forming them is, as above, $\mathbf{f}(1) = \mathbf{f}(2) = 1; \mathbf{f}(k+2) = \mathbf{f}(k+1) + \mathbf{f}(k)$ for $k \geq 1$.

The above commands can be put into an M-file. As a variant on the above we could use a 'while loop', as follows:

```
f=[1  1];
k=1;
  while f(k) < 1000
    f(k+2)=f(k+1)+f(k);
    k=k+1;
  end
f
plot(f)
```

This M-file is stored under the name `fibno.m` and is executed by typing `fibno`. The indentation used is intended to display the structure of the M-file and has no effect on the running or output. This time the two commands between `while` and `end` are executed while the condition, $\mathbf{f}(k) < 1000$, of the while loop is valid. Since $\mathbf{f}(16) = 987$ and $\mathbf{f}(17) = 1597$ the last value of k which allows execution of the while loop is $k = 16$. The value of k at the end of the whole M-file is 17 since k is incremented by 1 during the final execution of the loop. Notice an important difference between `for` and `while` loops:

- In a `for` loop the loop variable (k in the above example) is automatically incremented by 1 for each pass through the loop. If you want the variable incremented by d each time and starting at a while not exceeding b you use

  ```
  for k=a:d:b
  ....
  end
  ```

- In a `while` loop the variable (k above) is *not* incremented automatically. You have to increment it explicitly during the loop, as above where this is done with `k=k+1`.

3.2 A loop with conditionals: the $3n+1$ or hailstone problem

Let n be a positive integer. We iterate the following process: if n is *even* we divide it by 2; if n is *odd* we replace it by $3n+1$. Thus, starting with $n = 10$ we get successively the numbers 5,16,8,4,2,1. Here is an M-file which performs this automatically on any given (input) n. It is called `hail.m`. Note that in MATLAB `rem(a,b)` is the remainder on dividing a by b.

```
n=input('Enter n')
f=[n];
k=1;
  while n>1
    k=k+1;
    if rem(n,2)==0
      n=n/2;
    else
      n=3*n+1;
    end
  f(k)=n;
  end
bar(f)
```

To use this we type `hail` and then input say 25 <Enter>. This version produces a bar-chart rather than a graph.

The resulting bar-chart for $n = 1000$ is shown in Figure 3.1.

Typing k after running the above M-file produces an answer which is the exact number of steps taken to reach 1, counting the original number as Step 1. (Thus for example $n = 8$ produces $8, 4, 2, 1$ which counts as 4 steps.) For $n = 1000$ it takes 112 steps, as you can roughly see from the bar chart of Figure 3.1. Typing `max(f)` after running the M-file produces the maximum value attained by n during the iteration: the highest point on the bar-chart. This is 9232 for a starting value of 1000.

Here are some notes on the way that `hail.m` works.

(i) Note the incrementing of k in the `while` loop. This could equally well be done later in the loop, after the $f(k) = n$ line.

(ii) Note the `if...else...end` construction. All conditionals must be closed with an `end` statement. You can have `if...end` without an alternative 'else'.

(iii) The `==` symbol is used for comparison. The quickest way to

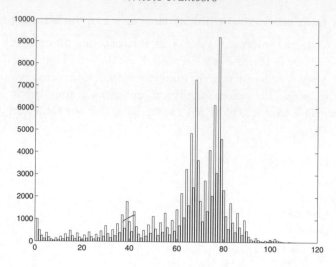

Fig. 3.1. The 'hailstone' iteration starting at $n = 1000$.

check whether something is even is to compute the remainder on division by 2, which is done by `rem(n,2)`.

(iv) Semi-colons after the `if` statement and the `else` statement are not needed (but will not do harm).

It is *not known* whether the above process *always* terminates in a 1 for any input n. No number has yet been found which does not terminate in 1 but no general proof exists either. You can read more in [13].

3.3 The euclidean algorithm for greatest common divisors

To find the greatest common divisor (gcd) of two whole numbers a and b, where $b > 0$, we first divide b into a:

$$a = bq + r, \text{where } 0 \le r < b.$$

Here q is the quotient and r is the remainder, which is calculated by

`r=rem(a,b)`

(If needed, the quotient q can be calculated by `q=floor(a/b)`.) Then we always have

$$\gcd(a,b) = \gcd(a - bq, b) = \gcd(r, b) = \gcd(b, r). \qquad (3.1)$$

Table 3.1. *Euclid's algorithm.*

Step	Current a	Current b	Current q	Current r
1	69	15	4	9
2	15	9	1	6
3	9	6	1	3
4	6	3	2	0

In the special case where r is an exact factor of b (denoted $r|b$ and pronounced 'r divides b'), the last term is simply r.

Sketch proof of (3.1) The first $=$ in equations (3.1) follows because the pair a, b have *exactly the same* divisors as the pair $a - bq, b$, hence certainly the same gcd. To see this, let $d|a$ and $d|b$. Then $a = a_1 d$, $b = b_1 d$ say for integers a_1, b_1. Then $a - bq = d(a_1 - b_1 q)$, so d is also a factor of $a - bq$. Conversely, the same kind of argument shows that a common factor of $a - bq$ and b is also a factor of a. The second $=$ in equations (3.1) is from the definition of r and the third $=$ just reverses the order of the numbers without affecting their gcd.

The idea of Euclid's algorithm is to replace a by b and b by r and repeat. Let us take $a = 69, b = 15$. The calculation can be organised as shown in Table 3.1. Then, $\gcd(69, 15) = \gcd(15, 9) = \gcd(9, 6) = \gcd(6, 3) = 3$ since, at the end, 3 goes exactly into 6, and hence the gcd is 3.

To realise this in MATLAB, we do the division and then replace a by b, b by q. The calculation is continued while the remainder r calculated is > 0, but there is some advantage in using the condition $b > 0$ since, at the end of the loop, they are equal anyway and r does not have a value until the loop has already been entered. It is much easier to understand the final M-file than to describe it! Can you see why the gcd is the final value of a calculated by the procedure, rather than the final value of b or r?

```
% Function for calculating the gcd of a and b.
% To use, type e.g. gcd(69,15)
% For convenience we make sure a,b nonnegative to start with

function h=gcdiv(a,b)
if a<0 a=-a;
end
```

```
if b<0 b=-b;
end
  while b > 0
    r=rem(a,b);
    a=b;
    b=r;
  end
  h=a;
```

This M-file, which is available as `gcdiv.m`, is written as a *function*. Note that the function value is h which is the outcome of the euclidean algorithm procedure. The M-file is used by a command of the form `gcdiv(a,b)` where a and b are either known to MATLAB at that moment, or are replaced by explicit whole numbers such as 69 and 15. What happens if you type `gcdiv(0,4)` ? Do you think the answer is reasonable? What about `gcdiv(4,0)` ?

There is an exercise to extend this to calculating the gcd h of *three* numbers, at the end of this chapter (Exercise 3.7), given the definition $\gcd(a, b, c) = \gcd(\gcd(a, b), c)$. That is, you work out the gcd x of a and b and then $h = \gcd(x, c)$. There is also an exercise to get the computer to generate a pair of *random* numbers for the calculation of the gcd. In the Investigations (Chapter 9) there is a project involving the gcds of large numbers of randomly chosen numbers. For this it is necessary to write an M-file which repeatedly uses the function `gcdiv` above.

3.4 Fermat's theorem and the power algorithm

The most interesting examples from elementary number theory use a wonderful theorem first stated in general by Pierre de Fermat in 1640, though the first full published proof is due to Leonhard Euler in 1736. A special case was known to Chinese mathematicians hundreds of years before Fermat. The theorem is proved in any book of elementary number theory, for example [7]. Here, it will merely be stated and some of its consequences explored. Recall that a prime number is a whole number $p > 1$ which has no factor besides 1 and p itself. The theorem is this:

Fermat's theorem *Let p be prime and suppose a is not divisible by p. Then $a^{p-1} - 1$ is divisible by p.*

The conclusion can also be written $\mathrm{rem}(a^{p-1}, p) = 1$, meaning 'the remainder on dividing a^{p-1} by p is 1', or $a^{p-1} \equiv 1 \bmod p$. We shall not make use of the notation $\equiv$ here; $a \equiv b \bmod m$, where a, b, m are whole numbers and $m \neq 0$, means simply that m is a factor of $a - b$.

We shall use Fermat's theorem shortly to give a method for showing when numbers are *not prime*, but for the moment we concentrate on the following:

Problem If n and m are reasonably large numbers, how can we calculate the remainder on dividing a^n by m?

If say $n = 1000000$, then even with $a = 2$ it is completely infeasible to work out a^n and then get the remainder from that, because $2^{1000000}$ has about 300000 digits! MATLAB fails at much smaller numbers, for example it correctly says that `rem(7^2,10) = 9` but it claims that `rem(7^20,10) = 0` which is obviously false!

3.4.1 The power algorithm

This is a cunning method for working out $\mathrm{rem}(a^n, m)$ when the numbers are large. The mathematical details are given in §3.5.1 for your interest. The main thing to remember is that the power algorithm does *not* work out a^n and only needs to handle numbers about as big as m^2 during the calculation. To use the power algorithm for working out $\mathrm{rem}(a^n, m)$, simply type

```
>> pow(a,n,m)
```

Remember that you can get information on the M-file `pow.m` by typing `help pow`.

Using the power algorithm it is easy to verify many examples of Fermat's theorem such as $16^{1998} \equiv 1 \bmod 1999$, which is so because 1999 is prime.

Examples

(i) `pow(3,118,119)` gives 32. Hence, by Fermat's theorem, 119 is *not* prime (for, if it were, then the answer would have been 1). Of course, $119 = 7 \times 17$.

(ii) `pow(2,1993002,1993003)` gives 1121689, so 1993003 is *not* prime (in fact, it is 997×1999).

(iii) `pow(2,10^6+2,10^6+3)` gives 1, so this is consistent with $10^6 + 3 = 1000003$ being prime, but doesn't *prove* it. Also

```
pow(3,10^6+2,10^6+3)
pow(5,10^6+2,10^6+3)
```

both give 1. All these are in some sense 'evidence' for $10^6 + 3$ being prime, but not a proof. (In fact, $10^6 + 3$ *is* prime.) See §3.4.2 for some nasty examples where the 'evidence' for primality is in fact false evidence!

Eventually the arithmetic in MATLAB starts to go wrong with rounding errors. For example, $10^9 + 7$ happens to be prime, but

```
>> pow(2,10^9+6,10^9+7)
```

does not give the correct answer 1. This is because ten figure numbers cannot be reliably manipulated by MATLAB. You can rely on `pow` for seven or eight figure numbers.

More powerful packages can handle hundreds of figures. These packages are routinely used to find very large (at least 100-digit) numbers which are 'probably prime', using Fermat's theorem and a more subtle variant called Miller's test. These 'probable primes' are used in cryptography, that is in sending messages by virtually unbreakable codes. You can read about such codes in [12]. Miller's test for more modestly sized numbers appears in Chapter 9.

3.4.2 Pseudoprimes

There are numbers which satisfy the *conclusion* of Fermat's theorem without satisfying the *hypothesis* that p is prime. That is, we can sometimes have $\text{rem}(a^{m-1}, m) = 1$, for some $a > 1$, without having m prime. When this happens we call m a *pseudoprime to base a*.

For example, $\text{rem}(7^{24}, 25) = 1$, $\text{rem}(2^{340}, 341) = 1$, so that $25 = 5^2$ is a pseudoprime to base 7, and $341 = 11 \times 31$ is a pseudoprime to base 2. Nevertheless if, for a fixed m, lots of values of a are found for which $\text{rem}(a^{m-1}, m) = 1$, then this can be regarded as accumulating evidence for the primality of m. For very large m this is by far the most efficient way of selecting 'probable' primes, though in practice a variant known as *Miller's test* is used. This is the subject of one of the Investigations. See Chapter 9.

The number 561 ($= 3 \times 11 \times 17$) has the rather unpleasant property that $\text{rem}(a^{560}, 561) = 1$ for *any a* not divisible by 3, 11 or 17. It is called a *Carmichael number*. Numbers such as these masquerade as primes under a large variety of tests, but can usually be unmasked with a little patience. It was proved only in 1992 that there are infinitely many Carmichael numbers.

Exercises

3.1 Run the M-file `fibno.m`, i.e. type `fibno` in MATLAB. The *length* of the vector `f` is obtained by `length(f)`. What is the length of `f` in this case? What is the largest Fibonacci number < 1000 and the smallest one > 1000?

3.2 Use the `hail` command to find the number of steps which 27 takes to reach 1, and the largest value reached during the iteration.

Modify `hail.m` so that it prints out the number of iterations and the largest value reached, but does *not* plot the values at the end. Call the resulting M-file `hail2.m`.

Modify `hail2.m` as follows, calling the result `hail3.m`. Remove the line `n = input('Enter n ');`, insert the lines

```
for k=50:100
n=k
```

at the beginning, and

```
end   % of for k=1:50 loop
```

at the end. What does this M-file do? Check this by running it, i.e. typing `hail3` in the MATLAB window.

Using a diary (§1.3.4) for storing the output from `hail3.m` read off all the values of n between 50 and 100 which have the same largest value during the hailstone iteration as 27 has. For each one, state the number of iterations each takes to reach 1.

3.3 Type `format short` <Enter>. Let 6_n and 7_n denote the whole numbers written in ordinary decimal notation as strings of n 6s and n 7s, respectively. So $6_4 = 6666$, etc. Use the command `gcdiv` to find the gcd of 6_n and 7_n for $n = 1, 2, 3, \ldots$ When is the displayed answer wrong? How do you know it is wrong?

Try typing `format long` and repeating the exercise. What is the new value of n? Again, how you can be sure that, for this n, the answer is wrong?

3.4   Typing pow(17,1000,100) should give the answer 1, so that
17^{1000} is 1 + a multiple of 100, i.e. $17^{1000} - 1$ is a multiple of
100. What are the last two digits of 17^{1000}? Show in a similar
way that $17^{1000} - 1$ is a multiple of 99 and 101. Why can you
deduce now that $17^{1000} - 1$ is a multiple of $99 \times 100 \times 101$? (This
uses the fact that no two of 99, 100 and 101 have a common
factor.)

Use pow to find the last three digits of 19^{2000}.

Find the remainders on dividing 17^{3313} and 3313^{17} by 112643.
Note: The MATLAB command rem is of *no use* here: the numbers are far too large. You must use the power algorithm M-file
pow. Writing $17 = 1 + 16$, why does it follow that $17^{3313} - 1$ is
divisible by 16? (That's a *mathematical* question!) Can you see
why, similarly, $3313^{17} - 1$ is divisible by 16? What can you say
about $\gcd(17^{3313} - 1, 3313^{17} - 1)$?

3.5   Use pow to find $\mathrm{rem}(2^{1394676}, 1394677)$. Why can you deduce
that 1394677 is *not* prime?

3.6   Type in an M-file as follows, calling it mypow.m

```
m=10^7+1;
while m<10^7+100
        m                  % Note no semicolon!
        pow(2,m-1,m)  % Note no semicolon!
        m=m+2;
end;
```

What does this do? Use it (and a diary if necessary, though
the pause button can be useful too) to find the only two numbers
between 10^7 and $10^7 + 100$ which could *possibly* be prime.

3.7   Create a function h=gcdiv3(a,b,c) which calculates the gcd of
three numbers a, b, c. Here is one way to do this. Your M-file
should start with the line
function h=gcdiv3(a,b,c)
and it should simply work out $x = \gcd(a, b)$ and then $h = \gcd(x, c)$, in each case by calling on the function gcdiv which
already exists. It is a general fact that

$$\gcd(a, b, c) = \gcd(\gcd(a, b), c)$$

so your M-file should be very short! Call the completed M-file
gcdiv3.m to indicate that it is an M-file.

Test the function `gcdiv3` on some triples of small numbers to see whether it gives the right answer. Then calculate

gcd(414304,56496,351824) and gcd(196054,175131,133407).

Create an M-file `gcdrand.m` with the following lines (which are one way of generating 'random' numbers, in this case numbers between 1 and 10^6)

```
t = fix(clock)
n = t(6)
rand('seed',n)
a = round(rand*10^6)
b = round(rand*10^6)
c = round(rand*10~6)
gcdiv3(a,b,c)
```

(We will be discussing 'random' numbers in more detail in Chapter 5.) Run this M-file 20 times and write down the gcds obtained (don't write down the numbers a, b, c). What percentage of trials give gcd equal to 1? There is an investigation later (Chapter 9) which finds the 'probability' that three randomly chosen numbers have gcd equal to 1, both theoretically and experimentally.

3.5 Appendix

3.5.1 Proof of the power algorithm

The key fact is that we can take remainders at any time during a calculation and always arrive at the same answer. Thus if we want $x = \text{rem}(ab, m)$ we can work out successively

$$p = \text{rem}(a, m), \quad q = \text{rem}(b, m), \quad x = \text{rem}(pq, m). \qquad (3.2)$$

For example rem(973 × 58,10) = rem(3 × 8, 10) = rem(24,10) = 4. This is of course just the *units digit* of 973 × 58, since we are taking remainders after division by 10.

There is an extremely fast algorithm for working out remainders of large powers. Here is an example. Suppose we want to work out $x = \text{rem}(7^{50}, 11)$. First, we work out the *binary representation* of the power, 50. This is

$$50 = (0 \times 1) + (1 \times 2) + (0 \times 2^2) + (0 \times 2^3) + (1 \times 2^4) + (1 \times 2^5).$$

Table 3.2. *The power algorithm.*

b	rem(b,11)	d=bit	x so far
7	7	0	1
7^2	5	1	5
7^4	3	0	5
7^8	9	0	5
7^{16}	4	1	9
7^{32}	5	1	1

Thus

$$7^{50} = 7^2 \times 7^{16} \times 7^{32}. \tag{3.3}$$

We'll see shortly how to do this quickly. Then in succession we work out the remainders on division by 11 of $7^2, 7^4, 7^8, 7^{16}, 7^{32}$. This is done by repeated *squaring* and using equations (3.2) successively. Thus we calculate rem(7^2, 11) explicitly, then square the answer (5) and take the remainder again. At that point we just get 3 and the square of *that* is 9—no need to take a remainder. The full sequence is:

$$\text{rem}(7^2, 11) = 5, \text{rem}(5^2, 11) = 3, 9, \text{rem}(9^2, 11) = 4, \text{rem}(4^2, 11) = 5.$$

Now using (3.3), we only need the first, fourth and fifth of these. It is best to set everything out as in Table 3.2. Note that the only time the 'x so far' entry changes is when there is a '1' in the 'bit' column, indicating that that power of 2 does indeed occur in the expression (3.3).

The way that the bits are calculated is as follows. We start with $d = \text{rem}(n, 2)$ as the 'units bit' for n, that is, the right-most bit in the binary expansion of n. At every step after the first, n is replaced by $(n - d)/2$; this brings the next bit in the binary expansion to the right-most position. Thus for 50 we get successively:

d=rem(50,2)=0; $n = (50 - 0)/2 = 25$;
d=rem(25,2)=1; $n = (25 - 1)/2 = 12$;
d=rem(12,2)=0; $n = (12 - 0)/2 = 6$;
d=rem(6,2)=0; $n = (6 - 0)/2 = 3$;
d=rem(3,2)=1; $n = (3 - 1)/2 = 1$;
d=rem(1,2)=1.

The M-file `pow.m` to implement this is as follows:

```
function x=pow(a,n,m)
```

```
b = a;
x = 1;
  while n>0
    d = rem(n,2);
    if d==1
      x = rem(x*b,m);
    end
    b = rem(b * b,m);
    n = (n-d)/2;
  end
```

4

Graphs and Curves

In this chapter we shall use MATLAB to draw graphs of functions $y = f(x)$, to approximate functions in two different ways as polynomials, to solve equations $f(x) = 0$ and to draw systems of lines in the plane forming an 'envelope'. (For graphs $z = f(x,y)$, see Chapter 17.) The technique of approximation is particularly important in applications, where it is desirable to replace a relatively complicated function (or indeed data set; compare Chapter 5) by a simple polynomial which can be easily handled. The choice here is between a very good approximation of $f(x)$ close to some given value x_0 of x, which gets steadily worse as x moves away (Taylor polynomials), and a reasonably good approximation over a larger range ('polyfit' approximations). Each has its uses.

4.1 Polynomials

A polynomial such as $p(x) = x^4 + 2x^3 - 3x^2 + 4x + 5$ is entered by means of its coefficients:

```
>>  p=[1  2  -3  4  5]
```

Note the *spaces* between the numbers; commas are also allowed for separation. In other words, a polynomial is just entered as a *vector* containing the coefficients, starting with the 'leading' coefficient in front of the highest power of the variable x.

The *roots* of p (i.e., the solutions of $p(x) = 0$) are obtained by typing

```
>>  roots(p)
```

Notice that this finds the complex roots as well as the real ones. We can increase the number of figures displayed by typing

44

```
>>   format long
```

We can draw the graph of p by typing

```
>>   x=-4:.05:2;
>>   y=polyval(p,x);
>>   plot(x,y)
```

The first line creates a vector $\mathbf{x}$ whose entries are the numbers from 4 to 2 at 0.05 intervals. Thus

$$\mathbf{x} = [-4.0 \ -3.95 \ -3.9 \ldots 1.95 \ 2.0].$$

As usual, the semi-colon merely suppresses output to the screen.

The second line evaluates the polynomial p at all these $\mathbf{x}$, producing a vector $\mathbf{y}$ of the same length as $\mathbf{x}$ containing $p(-4), \ldots, p(2)$.

The third line plots values of $\mathbf{y}$ against values of $\mathbf{x}$, taking them in corresponding pairs and joining up with (very short) straight segments.

By adding

```
>>   hold on
>>   plot([-4,2],[0,0])
```

the x-axis is also plotted. The plot command takes the first entry from each square bracket, making the point $(-4, 0)$, and joins it to the point $(2, 0)$ obtained from the second entry from each bracket, thus drawing the x-axis across the screen. From this it is quite clear that p has two real roots in the interval from -4 to 2, and their positions agree with the calculation of roots as $-3.18\ldots$ and $-0.728\ldots$. See Figure 4.1.

4.2 Initial examples of drawing curves

First recall the basic steps involved in plotting (§1.2.2). For example, to draw the curve $y = \sin x$ for x between 0 and 2π we type

```
>>   x=0:.05:2*pi;
>>   y=sin(x);
>>   plot(x,y)
```

The command `plot` works with two (identical length) vectors of values specifying pairs of coordinates. In this case, if you type `length(x)` or `length(y)` you should get the response **126**. Although axes are scaled automatically by the maximum values of x and y, they can be manually changed if necessary. For example, the line

Graphs and Curves

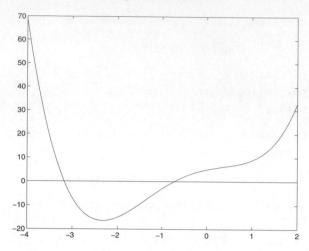

Fig. 4.1. Graph of $y = x^4 + 2x^3 - 3x^3 + 4x + 5$, with the x-axis also drawn.

```
>>   axis([0 10 -2 2])
```

makes the screen cover the region $0 \le x \le 10, -2 \le y \le 2$. Try this to see the effect. (In versions of MATLAB before Version 4, you may need to reissue the `plot(x,y)` command after `axis`.)

Similarly, to plot the graph of $y = x \sin x$, say, we could use

```
>>   x=0:.05:2*pi;
>>   y=x.*sin(x);
>>   plot(x,y)
```

Note the `.*` operation (§1.2.1), which multiplies corresponding entries of the vectors `x` and `sin(x)`, and does not attempt the impossible task of multiplying `x` and `sin(x)` together as matrices (they're the wrong size).

Now try typing

```
>>   x=0.01:.01:1;
>>   y=1/x;
>>   y=1./x;
>>   plot(x,y)
```

The first attempt to define $y = 1/x$ will not work; the second *should* produce the graph $y = 1/x$, for $-1 \le x \le 1$. Recall (§1.2.1) that the symbol `./` is used for elementwise division: divide 1 by each of the elements of the vector **x** in turn and call the resulting vector **y**. The instruction `1/x` makes no sense since you can't divide 1 by a *vector*. (In versions of MATLAB before Version 4, `y=1./x` was not acceptable either and a *space* had to be left after the 1 so that the dot was not confused with a decimal point.)

Typing

```
>>   t=0:.05:2*pi;
>>   x=cos(t);
>>   y=sin(t);
>>   plot(x,y)
```

produces the parametric curve $\{(\cos t, \sin t)\}$, which is of course a circle of radius 1 centred at (0,0), although it won't look very circular because of the automatic scaling. Try

```
>>   axis('square')
```

and note how the plot changes to being circular. You can also use

```
>>   axis('equal')
```

which makes the scales on the two axes equal no matter what the shape of the figure. Thus

```
>>   t=0:.05:2*pi;
>>   x=2*cos(t);
>>   y=sin(t);
>>   plot(x,y)
>>   axis('equal')
```

produces a picture of an ellipse where the major and minor axes have their true ratio of 2:1. Replacing `axis('equal')` with `axis('square')` does not have the same effect. Try it!

There are parametric curves in the exercises and investigations which are more exciting than this!

4.3 Taylor polynomials

A good illustration of the power of graphics is to compare the graph of a function such as $\sin x$ and the graph of its Taylor polynomial approximation

$$x - \frac{x^3}{3!} + \frac{x^5}{5!} - \ldots + (-1)^{k-1} \frac{x^{2k-1}}{(2k-1)!}.$$

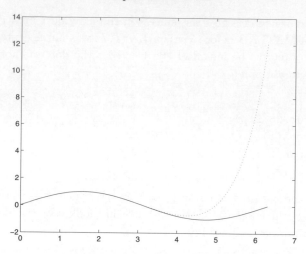

Fig. 4.2. Sine curve (solid) and Taylor approximation of degree 9 (dotted).

The M-file `tsine.m` calls for the value of k (this is the *number of terms* in the polynomial), then plots the 'true' curve in green and, once <Enter> is pressed, plots both this and the approximation in red. MATLAB uses first the scale appropriate to the sine curve alone and then the scale appropriate to the two curves together. Remember that to run this M-file you just type `tsine` followed by <Enter>. You never type the `.m` when running M-files.

In Figure 4.2 the sine curve is the solid line and the Taylor approximation with $k = 5$ (that is, degree $2k - 1 = 9$) is the dotted line.

The M-file `tsine2.m` on the other hand plots the actual sine curve (in green) and then *freezes the scaling* while the approximation is plotted (in red). This is achieved by the command `axis(axis)` after the first plot command in the M-file. Notice the difference with say $k = 4$.

The approximation hugs the true curve more and more closely from $x = 0$ onwards, as k increases. By the time k reaches about 10, the curves are more or less indistinguishable over the range from 0 to 2π.

Here is the text of the M-file `tsine`. Even if you don't follow all the details, you should note how *easy* it is to do this relatively complicated thing in MATLAB.

```
% Draws the Taylor approximation
% to a sine curve of degree 2k-1
k=input('Type the number of terms
```

```
              of the Taylor series for sine:   ');
x=0:.05:2*pi;
z=sin(x);
plot(x,z,'g') %Actual sine curve is drawn in green
hold on
pause % This holds up execution until Enter is pressed
        w=x;
        y=x;
        s=-1;
        for j=1:k-1
                w=w.*x.*x/(2*j*(2*j+1));
                y=y+s*w;
                s=-s;
        end
plot(x,y,'r')
%Approximation is drawn in red
hold off
```

Note the use of 'g' to draw in green (see §1.2.2). A dashed curve is drawn by plot(x,y,'- -'); when needed, the symbol - denotes a solid curve. We have also already met the hold on statement which holds the graph and puts the second one on top of it. It is necessary to 'release' the graphs by a call to hold off at the end. Notice again the use of elementwise multiplication .* in this file.

4.4 Approximations using the function polyfit

An entirely different approach to approximating a function by a polynomial is taken by the available function polyfit, which fits a polynomial of a given degree to a curve by a method which in a sense minimises the distance of the polynomial graph from the true curve graph, averaged along their whole lengths. (See also Chapter 5.) For example (this M-file is called polyex.m):

```
a=0:.05:2*pi;
b=sin(a);
c=polyfit(a,b,5);
d=polyval(c,a);
plot(a,d,'r')
hold on
pause
```

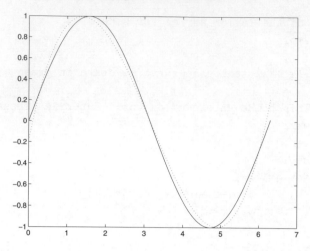

Fig. 4.3. Sine curve (solid) and 'polyfit' approximation of degree 3 (dotted).

```
plot(a,b,'g')
hold off
```

This finds a polynomial approximation of degree 5 to the sine curve over the range 0 – 6.3, using a for a change as the independent variable. The polynomial approximation c is simply a list of *coefficients*; to evaluate it at the points of a we use the function `polyval`. The first 'plot' command draws the approximation in red and the second 'plot' command draws the true sine curve in green. Notice how completely different this approximation is from the Taylor approximation of the same degree (given by $k = 3$ in the M-file `tsine.m`).

Figure 4.3 shows the result of running `polyex.m` with degree 3.

4.5 The goat problem

Here is a well-known problem which leads to a frightful equation. We can solve it graphically, or numerically.

Problem A goat is tethered to a point on the circumference of a circular field of radius 10 metres. How long should the rope be so that the goat can graze over exactly *half* the area of the field?

Let x be the length of the rope in metres. (That was the easy bit!) Some messing around with areas of sectors and areas of triangles gives the following equation for x:

$$2(x^2 - 200)\cos^{-1}\left(\frac{x}{20}\right) - x\sqrt{400 - x^2} + 100\pi = 0. \qquad (4.1)$$

See Figure 4.4 and §4.7.1. Note that $\cos^{-1}$ is the inverse cosine, which in MATLAB is `acos`. It does not seem very feasible to solve this equation explicitly! For a graphical solution the left-hand side of (4.1) can easily be plotted—the M-file `goatgr.m` does this. It should be clear from the plot (also perhaps from looking at Figure 4.4) that there is a solution *somewhere* around $x = 10$. One way of getting MATLAB to calculate an accurate value of the solution is to use the built-in function `fzero`. For this we need an M-file containing a definition of the function in question. In this case it is `goatfn.m` and is as follows (this function is available to you; no need to type it in!):

```
function y=goatfn(x)
    y=2*(x.*x-200).*acos(x/20)-x.*sqrt(400-x.*x) + 100*pi;
    end;
```

Note that the MATLAB command `goatfn` will produce an error, but say `goatfn(10)` will give the value of the function when $x = 10$. What happens when you type `goatfn(25)` ? What in fact is the 'domain' of the goat function, that is, the range of values of x for which y is real?

We now type

```
>>    a=fzero('goatfn',10)
```

meaning 'look for the solution a of goatfn$(x)=0$ near to $x = 10$'. By using `format long` a *very* accurate estimate is found.

4.6 Envelopes of lines

There is an attractive way of using the graphics in MATLAB to draw systems of lines in the plane and their *envelopes*. For instance the M-file `parnorm.m` draws the normals to the parabola $y = x^2$, parametrised by $x = t, y = t^2$. The normal to the parabola at (t, t^2) has the equation

$$y - t^2 = -\frac{1}{2t}(x - t), \quad \text{that is, } x + 2ty - t - 2t^3 = 0.$$

This is $p(t)x + q(t)y + r(t) = 0$, where $p(t) = 1, q(t) = 2t$ and $r(t) = -t - 2t^3$. So the functions p, q, r, written in MATLAB, are

```
p=ones(size(t));
```

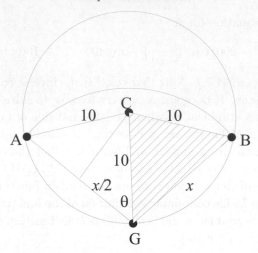

Fig. 4.4. A goat is tethered at G. We want to discover the length x of rope which will allow the goat to graze over half the field. The resulting equation for x is (4.1), and this is proved in §4.7.1. The shading is for that proof.

```
q=2*t;
r=-t-2*t.*t.*t;
```

Note the use of `ones(size(t))` when the function is simply a constant equal to 1. If it had been $p(t) = 5$ say, then we would use `p=5*ones(size(t))`. Note also the use of `.*` for cubing each value of t. The form `r=-t-2*t.^3` also works.

Other possibilities for p, q, r are considered below.

The M-file `parnorm.m` requires input of

   (i) the bounds on x (use $xl = -2, xu = 2$, that is we are restricting the x-coordinate to lie between -2 and 2);

  (ii) the lower limit of y (use $yl = -1$). The upper limit of y is then determined if we require that the range of x, namely 4, equals the range of y;

 (iii) the bounds on the parameter t (use $tl = -2, tu = 2$, so that t goes from -2 to 2, like x).

When you run `parnorm.m` (type `parnorm` and <Enter>) the effect is to draw all the lines on the screen and the eye irresistibly picks out another curve with a downwards cusp, which is *tangent* to all the lines. This cusped curve is called the *envelope* of the lines. Using <Enter> after

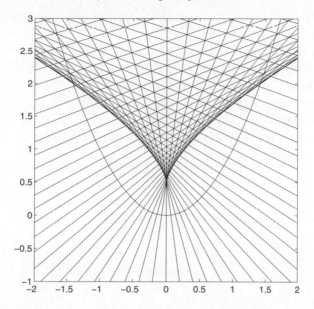

Fig. 4.5. The normals to the parabola $y = x^2$ and their 'envelope', a downward cusped curve.

the lines are drawn actually emphasises this cusped curve by drawing it in white. See Figure 4.5.

The M-file `parnorm.m` is a special case of the M-file `linenv.m`, which draws a family of lines

$$p(t)x + q(t)y + r(t) = 0,$$

where p, q, r are any functions of a parameter t, over some pre-defined range of values of t. Thus we obtain the M-file `parnorm.m` from `linenv.m` by specifying the functions p, q, r in the correct place (see below). It has been adapted to draw the parabola in red and (after pressing <Enter>) the envelope curve itself in white, by adding a few lines to the file. See Exercise 4.7 for a method of calculating the envelope curve. You can read more about envelopes in Chapter 10.

The general M-file `linenv.m` is **incomplete**: before it will run you need to specify the functions p, q, r, in the following place:

```
p=              % These three lines are where the
q=              % equation of the line, px+qy+r=0,
```

```
    r=                      % is placed. p,q,r are functions of t.

% BEWARE!! that say q=4 will not produce a
% vector but a constant
% number.  So if any of p,q,r is
% constant, say k, then you need
% to write it as k*ones(size(t)).
% Here, ones((size(t)) is a vector of 1's of
% the same length as the vector t.
```

In the particular case of the normals to a parabola (as in `parnorm.m`) we need the equation of the normal to the curve $y = x^2$ at the point (t, t^2), which is calculated at the beginning of this section.

Exercises

4.1   Try

```
    >>   a=1;
    >>   x=-5:.05:5;
    >>   y=x.*x.*x - a*x + 1;
    >>   plot(x,y)
```

Despite the fact that 1 is a number, MATLAB is sufficiently intelligent to add it to the vector `x.*x.*x` whose entries are the values of x^3 for $x = -5, x = -4.95, x = -4.9, \ldots, x = 5$. Thus 1 is added to each entry in this vector.

Can you tell from the graph how many real roots the equation $x^3 - x + 1 = 0$ has? Try

```
    >>   axis([-5 5 -3 3])
```

This should help: it forces MATLAB to adopt $-5 \le x \le 5$ and $-3 \le y \le 3$ on the axes.

The M-file `cubics.m` does the above, allowing you to choose your range of values of x and y, and also the value of a. So `cubics` <Enter> followed by, in succession, the five numbers $-5, 5, -3, 3, 1$, each followed by <Enter>, plots $y = x^3 - ax + 1$ with $-5 \le x \le 5, -3 \le y \le 3$, and putting $a = 1$. It also draws the x-axis in white. Look at the M-file if you want to see how it does this.

Try running `cubics` with other values of a and find approximately the value of a at which the graph *is tangent to* the x-axis, so that there is a *double root*. Once you have a fairly good idea of the value of a, run `cubics` with $0.5 \leq x \leq 1$ and $-0.05 \leq y \leq 0.05$. You'll find you can tell very accurately whether the graph is tangent to the axis. Find the best approximation you can for a.

4.2   In the M-file `tsine.m`, change
`plot(x,z,'g')` to `plot(x,z.*z,'g')`, and
`plot(x,y,'r')` to `plot(x,y.*y,'r')`.
Call the resulting M-file say `tsine3.m`. Which functions are now being plotted in red and in green? Give formulae for them. Run the M-file to discover the least number of terms (k) you need in order to get a good polynomial approximation for $0 \leq x \leq 4$. What is the *degree* of the resulting polynomial approximation? (*Hint*: It's *not* k.)

4.3   For the 'goat' problem (§4.5), find the value of x close to 10 using `fzero`. Is there just one possible answer to the original problem? (Look at the graph of the goat function.)

Try also the following, which approximates the goat function y by a polynomial z of degree 3 (you could do this by direct input to MATLAB or via an M-file):

```
x=-19:.05:19;
y=goatfn(x);
z=polyfit(x,y,3);
v=polyval(z,x);
plot(x,y,'g',x,v,'r')
```

This plots the true goat function in green and the approximation in red. You should find that the two curves are very close over the range from –19 to 19.

Type z to find out what the polynomial z is. Use `roots(z)` to find its roots. Why is there an 'extra' root for z which is not related to a zero of the goat function?

4.4   Type `x=.05:.01:2;` and then type in the necessary lines to draw the graph of $y = x \sin \frac{1}{x}$, over the range $0.05 \leq x \leq 2$. (Beware of that innocent looking $\frac{1}{x}$. Compare §4.2 above.) Use a sequence of commands similar to that in Exercise 4.3 to plot a polynomial approximation of degree 7 on the same diagram as the original curve. Why do you think the fit is so bad?

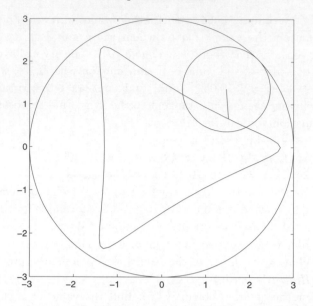

Fig. 4.6. Generation of a hypocycloid by a circle radius b rolling inside a circle radius $a > b$. The hypocycloid is generated by a point P fixed to a radius CQ of the moving circle, where the length of CP is d. In the figure, $a = 3, b = 1, d = 0.7$.

4.5   The M-file `hypocy.m` draws so-called *hypocycloid* curves (also known as *spirographs*), which are obtained by rolling a circle of radius b inside a circle of radius $a > b$. The curve is traced out by a point P rigidly attached to the rolling circle, at a distance d from its centre.

The parametrisation of the hypocycloid is

$$x = (a-b)\cos\left(\frac{bt}{a-b}\right) + d\cos t, \quad y = (a-b)\sin\left(\frac{bt}{a-b}\right) - d\sin t.$$
(4.2)

See the §4.7.2 and Figure 4.6. The M-file calls for the values of a, b, d and for the upper limit of t, that is, $0 \le t \le n\pi$, where n is input by you. Draw the curve using `hypocy.m` for $b = d = 1$ and $a = 2$. Explain what you see, using the above parametrisation.

Take $a = 3, 4, 5, 6, 7$, with $b = d = 1$ and find in each case: (i) the smallest integral value of n which makes the curve close up and start to repeat; (ii) the number of 'cusps' (sharp points) on the curve.

Can you formulate the general results here, for any integer $a > 2$?

4.6   Show that the perpendicular bisector of the line joining (a, b) to (c, d) has equation

$$2x(a - c) + 2y(b - d) - a^2 - b^2 + c^2 + d^2 = 0.$$

(*Hint*: Use the *equal distance* property: (x, y) is on the perpendicular bisector if and only if its distances from (a, b) and (c, d) are equal. This gives

$$(x - a)^2 + (y - b)^2 = (x - c)^2 + (y - d)^2.)$$

Deduce that the perpendicular bisector of the line joining $(0, 1)$ to $(t, 0)$ has equation

$$2xt - 2y + 1 - t^2 = 0. \tag{4.3}$$

Complete the M-file `linenv.m` to plot the envelope of these perpendicular bisectors for $-2 \le t \le 2$.

4.7   Suppose that the equations

$$px + qy + r = 0 \quad \text{and} \quad p'x + q'y + r' = 0 \tag{4.4}$$

(where p, q, r are functions of t as in the above examples, and $'$ denotes differentiation with respect to t) are solved for x and y:

$$x = \frac{r'q - rq'}{pq' - p'q}, \quad y = \frac{rp' - r'p}{pq' - p'q}. \tag{4.5}$$

(This is just solving two simultaneous linear equations.) Thus x and y are functions of the parameter t; as t varies the point (x, y) traces out a curve.

A straightforward but tedious calculation shows that:

The tangent to this curve (x, y) at the point with parameter t is precisely the original line $p(t)x + q(t)y + r(t) = 0$.

Thus (4.5) gives the envelope curve, which is tangent to all the lines.

In practice it is usually simpler to solve directly rather than use (4.5). For example, with the normals to a parabola in §4.6, we get the envelope from

$$x + 2ty - t - 2t^3 = 0 \quad \text{and} \quad 2y - 1 - 6t^2 = 0,$$

where the second equation is obtained by differentiating the first with respect to t. The second equation gives y directly, and then we get x from the first equation:

$$x = -4t^3, \quad y = \frac{1}{2}(1 + 6t^2).$$

This is the downward cusped curve in Figure 4.5. We can find the equation of the curve in x, y coordinates by eliminating t. Thus $6t^2 = 2y - 1$ so $216t^6 = (2y - 1)^3$. Also $4t^3 = -x$ so $16t^6 = x^2$. Hence $216x^2 = 16(2y - 1)^3$, that is,

$$27x^2 = 2(2y - 1)^3$$

is the equation of the envelope curve in this case.

Find the equation of the envelope in the case of the perpendicular bisectors example (4.3). How does this relate to the shape of the curve which appears when all these perpendicular bisectors are drawn on the screen?

4.7 Appendix

4.7.1 Proof of the goat equation

Referring to Figure 4.4, the region the goat, tethered at G, can graze is enclosed by a circular arc from G along the boundary of the field to A, then along a circular arc centred at G to B and finally along the boundary of the field back to G. The area of this region is the sum of:

(i) the area of a circular sector of the circle centre G, radius x, bounded by the straight lines BG, GA and the arc centred at G from A to B;

(ii) twice the area of the piece of the field below the chord BG.

The area of the latter piece is the area of a circular sector of the field from C along the radius to B, along the circumference to G and along the radius to C, *minus* the area of the shaded triangle.

Let θ be the angle CGA (equal to the angle CGB). The area (i) is $\frac{1}{2}x^2(2\theta)$ and $\theta = \cos^{-1}\frac{x}{20}$, so the area is $x^2 \cos^{-1}\frac{x}{20}$. The area of the piece in (ii) is

$$\frac{1}{2} \times 10^2 \times 2\left(\frac{\pi}{2} - \theta\right) - \frac{1}{2}x\sqrt{100 - \left(\frac{1}{2}x\right)^2}.$$

Adding (i) to twice (ii) and putting this equal to half the area of the circular field, 50π, gives equation (4.1).

4.7.2 Proof of the parametrisation (4.2)

Using the notation of Figure 4.6, the centre C of the rolling circle has reached an angle u anticlockwise from the horizontal, so that

$$C = ((a - b)\cos u, (a - b)\sin u).$$

At the same moment, the line joining C to the moving point P has rotated an angle t clockwise from its initial horizontal position. Here, we think of the small circle starting off with C on the positive x-axis, P starting at P_0. The *rolling condition* is that the arc of the large circle from P_0 to A equals the arc of the small circle from A to Q, where CPQ is a straight line as in the figure. Hence

$$b(u + t) = au, \quad \text{so that } u = \frac{bt}{a - b}.$$

The formula (4.2) follows because the point P is

$$((a - b)\cos u, (a - b)\sin u) + d(\cos t, -\sin t);$$

the last $-$ sign occurs because t is measured clockwise.

5

Representation of Data

As technologies advance, the amount of data becoming available in any practical application area is increasing rapidly. The purpose of representing data in another form is to help make use of this mass of data and to extract further information from it in a useful way.

In this chapter we present an introduction to data analysis using some of the elementary statistical tools within MATLAB. We consider characteristics within a data set, relationships between data sets and the extraction of particular data from data sets. Thus the topics we cover here coincide with the elementary functions of many modern statistical, spreadsheet and database packages.

5.1 Data analysis

Given a set of data $\mathbf{D} = [d_1, d_2, \ldots, d_n]$, we can compute its simple statistics such as maximum, minimum, mean and median (the value with 50% of the data above it and 50% below it). The corresponding MATLAB commands are `max, min, mean, median`. Here $\mathbf{D}$ can also be a column vector or a matrix.

To introduce some more commonly used commands, let us take

```
>> D = [2 1 3 4 7]
```

as an example. Then Table 5.1 shows how the statistics† of this data set

---

† If we denote the mean of $\mathbf{D}$ by $\bar{d}$, then a measure of the dispersion or variability is given by the *standard deviation* s with

$$s^2 = \frac{1}{n-1} \sum_{i=1}^{n} (d_i - \bar{d})^2 = \frac{1}{n-1} \left(\sum_{i=1}^{n} d_i^2 - n\bar{d}^2 \right).$$

Table 5.1. *Computing the statistics of a data set (vector* **D***).*

Description	MATLAB input line			Result
Minimum	m	=	min(D)	1
Maximum	M	=	max(D)	7
Mean	p	=	mean(D)	3.4
Median	P	=	median(D)	3
Product of elements in D	pr	=	prod(D)	168
Sum of elements in D	dsum	=	sum(D)	17
Cumulative Sum	csum	=	cumsum(D)	[2 3 6 10 17]
Standard deviation	s	=	std(D)	2.3022
Sorting D	so	=	sort(D)	[1 2 3 4 7]

can be calculated with MATLAB. Here note that *sorting* is only done in the ascending order; so what result would the following produce?

```
>>   D1 = - sort(-D)
```

A useful aid for displaying data is the *histogram*. The MATLAB command is simply `hist`; for example, `hist(`**D**`)` will display the *histogram* of the data set **D**. By default, the interval $\min(\mathbf{D}) \leq x \leq \max(\mathbf{D})$ is partitioned into ten equally spaced subintervals and the height of the bars is the number of data in the subinterval. We can vary the number of bars; to have three subintervals, type

```
>>   hist(D, 3)
```

So, here, the intervals are $1 \leq x \leq 3$, $3 < x \leq 5$, $5 < x \leq 7$ and the bars are centred on 2, 4, 6 respectively. To find the sizes (**y**) of the groups, type

```
>>   y = hist(D, 3)
```

and to find the sizes (**y**) of the groups and mid-points (**x**) of the intervals, type

```
>>   [y x] = hist(D, 3)
```

Here the results are **y** = [3 1 1] and **x** = [2 4 6].

There is a link with bar charts: type `help bar` to find details on using `bar` and you will see that `bar(x, y)` produces an identical plot to `hist(`**D**`, 3)`.

5.1.1 *Sorting*

Often it is necessary to *sort* a number of data sets so that one of them is in increasing or decreasing order. For instance, when there are two related data sets, the second set must be reordered consistently with the first set.

An example of this is in the M-file `tomato.m`, which compares the amount x of fertiliser applied to a tomato plant with the weight y of tomatoes produced.

Typing `tomato` produces two vectors **x** and **y** but the entries in **x** are not in increasing order. (You can see the vector **x** by typing **x**, of course.) Typing `plot(x,y,'g*')` produces a *scatter plot* of the data, x against y, the symbol used being green '*'.

Typing `plot(x,y)` produces a line plot, but because the data are not ordered it appears as a scribble rather than a graph. One way of making a proper plot is by typing

```
>>   [sx k] = sort(x);
>>   sy = y(k);
>>   plot(sx, sy)
```

The first line sorts **x** into increasing order, noting the positions of their components in the original order in the vector **k**. The sorted **x** are placed in a vector **sx**. The second line reorders **y** so that their components correspond to those of **x** correctly. Type [**x y sx sy**] to see the four columns of figures at once.

You will find that the line plot now looks more like a graph. Finally, as we mentioned earlier in this chapter, it is easy to sort data into a decreasing order with **sort** by taking the negative of the values.

5.1.2 *Querying*

Querying or extracting data lies at the heart of databases, now increasingly used in various management information systems. Here we shall take two simple examples to illustrate. The essential quantity to specify in using *query* is a *condition* often called the criterion. In MATLAB, this condition can be represented by a vector (or a matrix) with 0s and 1s, meaning 'not satisfied' and 'satisfied' respectively.

As a first example, consider those tomato plants which give more fruit. The task is to show a line plot of fertiliser versus weight for those with over 5 kg of fruit. To identify such cases out of the entire data set, we use a vector **d** with 1s and 0s, count the total number t of the 1s, and extract the satisfied cases in new vectors **x1** and **y1** (both sized t):

```
>>   tomato;
>>   d = (y>=5);     % Query Condition !!!!
>>   t = sum(d)      % Number of 'satisfied'
>>   [v k]=sort(-d); % Bring 'satisfied entries' to start
>>   k = k(1:t)      % Shorten the index to 't' (Query)
>>   x1=x(k); y1=y(k); % Pick up query results from x,y !
```

If we now wish to plot **x1** versus **y1**, a sorting of both according to **x1** has to be done first, as we discussed before. Here $t = 7$.

The second example is a simple geometric task. Given a set of 42 points (x, y) on a unit circle, we are to identify those points which lie in the second quadrant, and plot them:

```
>>   format compact; t=0: 0.15: 2*pi; n_count=length(t)
>>   x=cos(t);  y=sin(t);  % ------ the given data set---
>>   d = (y>=0) & (x<=0);  % Query Conditions !
>>   t = sum(d)            % Number 'satisfied'
>>   [v i]=sort(-d);       % sort 'satisfied entries'
>>   i = i(1:t)            % Shorten index to 't' (Query)
>>   x1=x(i); y1=y(i);     % Pick up query results.
>>   plot(x1,y1)           % ------ plot them -----------
```

Try it. Did you find $t = 10$?

5.2 Least squares fitting

Given a pair of data sets (say **x** and **y** in the 'tomato' example above) we may infer, by inspecting their scatter plot, a relationship or association between them. We want to find a function which approximately describes such a relationship. Of course, different choices of functions can be tried for such a purpose depending on what the scatter plot looks like. Suppose that such a pair of data sets is given by (x_i, y_i) for $i = 1, 2, \ldots, n$. We can conveniently write these as two vectors:

$$\mathbf{x} = [x_1, x_2, \ldots, x_n]^\top,$$

$$\mathbf{y} = [y_1, y_2, \ldots, y_n]^\top.$$

Here, we have taken them to be *column* vectors.

5.2.1 Fitting a straight line (the least squares line)

Perhaps the easiest function to try is the linear function

$$Y = \alpha + \beta x,$$

that is, we try to find two scalars α, β such that the linear relationship

$$\begin{cases} Y_1 = \alpha + \beta x_1, \\ Y_2 = \alpha + \beta x_2, \\ \quad \vdots \\ Y_n = \alpha + \beta x_n, \end{cases} \quad \text{that is,} \quad \begin{pmatrix} Y_1 \\ Y_2 \\ \vdots \\ Y_n \end{pmatrix} = \begin{pmatrix} 1 & x_1 \\ 1 & x_2 \\ \vdots & \vdots \\ 1 & x_n \end{pmatrix} \begin{pmatrix} \alpha \\ \beta \end{pmatrix}$$

produces $Y_i = y_i$ for all i. We denote this linear system by $Ac = \mathbf{Y}$ with $\mathbf{c}^\top = (\alpha\ \beta)$. Do such scalars exist? This is an over-determined system of linear equations for unknown quantities α and β which means that the system cannot be satisfied exactly,† in general, for any α and β.

An arbitrary choice of α and β would give an error of size $|y_i - Y_i|$ for each i, and trying to satisfy the system approximately implies a compromise with such a residual error. The Gaussian method of *least squares* says that‡ we should find that unique solution α and β which minimizes the sum of residual squares

$$R^2 = \sum_{i=1}^{n}(y_i - Y_i)^2,$$

where for all i,

$$Y_i = \alpha + \beta x_i.$$

5.2.1.1 First method: using \

In MATLAB, given data sets $\mathbf{x}$ and $\mathbf{y}$, we can find the best solution α and β easily by doing the following

```
>>  A=[ones(size(x)) x]
>>  c=A\y
```

Here, A is the matrix appearing above, with 1s down its first column and $\mathbf{x}$ down its second column. The second line instructs MATLAB to find the best approximation $\mathbf{c}$ to a solution of the equations $Ac = \mathbf{y}$. As defined before, $\mathbf{c}(1) = \alpha$ and $\mathbf{c}(2) = \beta$.

---

† Mathematically the solvability depends on whether vector $\mathbf{Y}$ lies in the span of the two column vectors of A! As $rank(A) \leq 2$ and often in practice $n > 2$, the system is more likely to be inconsistent, that is, have no solution!
‡ Further details are given in an appendix to this chapter.

Once α and β are available, we may use the relationship

$$Y = \alpha + \beta x$$

to find any points on the best-fitting line; for example, the following

```
>>   Y = c(1) + c(2)*x;
>>   plot(x,Y)
```

plots the best-fitting line (the least squares line).

To find the residual R^2 of the least squares fit, type (noting that **y, Y, r** are *column* vectors)

```
>>   Y = A*c ;
>>   r = y - Y ;
>>   R2 = r'*r
```

5.2.1.2 Second method: using polyfit

Given data as above, type

```
>>   C = polyfit(x, y, 1)
```

to find the best-fitting curve of degree 1 (that is, a line). This gives $\mathbf{C}(2) = \alpha$ and $\mathbf{C}(1) = \beta$. *Notice that* **c** *was a column vector while* **C** *is a row vector!* To find the residual of the least squares fit using the row vector **C**, type

```
>>   Y = polyval(C,x);
>>   r = y - Y;
>>   R2 = r'*r
```

Remember from §4.1 that `polyval` evaluates the polynomial C at the values of x in the vector **x**. That is, it finds the points on the best-fitting line. Since **x, y** are column vectors, so are **Y, r**.

5.2.1.3 Example data: tomatoes again

The data have been described in §5.1.1. Some information can also be obtained by typing `help tomato`. You should find the least squares solution is $\alpha = 4.3985$ and $\beta = 0.0966$, and the residual $R^2 = 0.388$.

You can fit and then plot a straight line without sorting the data, but if you want to fit and then plot any other curve (see §5.2.2) then you *do* need to sort the data.

5.2.2 *Fitting a polynomial function*

If the relationship between data sets **x** and **y** is not really linear, we may try nonlinear functions. The simplest example is the use of a quadratic function

$$Y = \beta_2 x^2 + \beta_1 x + \alpha.$$

As before, we want to find coefficients β_2, β_1 and α such that the following residual is minimised

$$R^2 = \sum_{i=1}^{n} (y_i - Y_i)^2,$$

where for all i

$$Y_i = \beta_2 x_i^2 + \beta_1 x_i + \alpha.$$

With MATLAB simply type

```
>>  C = polyfit(x, y, 2)
```

to obtain the coefficients $\mathbf{C}(1) = \beta_2$, $\mathbf{C}(2) = \beta_1$ and $\mathbf{C}(3) = \alpha$. To find the residual of the least squares fit using the row vector **C**, type

```
>>  Y = polyval(C,x);
>>  r = y - Y ;
>>  R2 = r'*r
```

Typing `help polyfit`, you will find out that `polyfit` can fit a polynomial function of any degree k to data sets **x** and **y**. The general function will be of the form

$$Y = \beta_k x^k + \cdots + \beta_2 x^2 + \beta_1 x + \alpha$$

and so the MATLAB command `C = polyfit(x, y, k)` gives the required coefficients in a row vector form

$$\mathbf{C} = [\beta_k, \ldots, \beta_2, \beta_1, \alpha].$$

Here the estimated values at all x_is can be found by

```
>>  Y = polyval(C,x);
```

So with `polyfit`, higher order polynomials can be considered with ease. But it cannot deal with the multiple variable case as discussed in §5.2.3.

For those who enjoy the beauty of the \ method, you will be pleased to know that this can be done. With the following hint on how to set up A, you should obtain the same solution as with `polyfit`

```
>>   A=[ones(size(x))   x   x.^2];   % .^ for component-wise
```

Remark: The nonlinear relationship between data sets **x** and **y** need not be polynomial; they may be other functions, for example, we can try the following

$$Y = \beta_2 e^{-x} + \beta_1 e^x + \alpha$$

or

$$Y = \beta_2 \sin x + \beta_1 \cos x + \alpha.$$

5.2.2.1 Example data: same tomatoes

Using the data sets in `tomato.m`, we can use `polyfit(x, y, 2)` to fit a quadratic function. The coefficients will be

$$[\beta_2 \quad \beta_1 \quad \alpha] = [-0.0045 \quad 0.1776 \quad 4.1826]$$

and the residual R^2 =0.217. Finally use `polyfit(x, y, 3)` to fit a cubic function. The coefficients will be

$$[\beta_3 \quad \beta_2 \quad \beta_1 \quad \alpha] = [0.0004 \quad -0.0144 \quad 0.2451 \quad 4.1088]$$

and the residual R^2 =0.1905.

Typing `toms` plots a linear fit, then a quadratic fit *without* sorting and finally a quadratic fit *with* sorting. (Hit <Enter> between plots to clear any pause.) Clearly sorting is vital for plotting quadratic fits! Look at the M-file `toms.m` to see how this is done, using sorting as in §5.1.1. Here is the relevant† part:

```
>>   [sx o] = sort(x);
>>   sy = y(o);
>>   C = polyfit(sx,sy,2);
>>   Y = polyval(C,sx);
>>   plot(sx,sy,'*w',sx,Y,'w')
```

5.2.3 Data fitting for multiple variables

The above methods assume that we have an 'output' dependent variable y which is a function of a single 'input' independent variable x. In the tomato example, the weight of tomatoes y is assumed to depend solely on the weight of fertiliser x; we have ignored sunlight and moisture

---

† Refer to Chapter 17 for plotting curved lines with sufficient plot points.

among other possible factors! It is often more realistic to assume that the 'output' data y is a function of several independent 'input' variables, and here we show how to handle this case on the assumption that the relationship is *linear* in all input variables. This assumption is only made to simplify the presentation; higher order polynomial relationships can be considered similarly. Thus, calling the 'input' variables $x^{(1)}, \ldots, x^{(p)}$, we are trying to find a relationship of the form

$$y = \alpha + \beta_1 x^{(1)} + \cdots + \beta_p x^{(p)}$$

in which $x^{(i)}$ occurs only to degree 1. In the example below, the response y to treatment of diabetics is assumed to depend on three factors, $x^{(1)} =$ age, $x^{(2)} =$ weight and $x^{(3)} =$ diet (which is converted into a numerical value according to some scheme).

We are often dealing with discrete data sets and so we assume p variables (or 'input' factors) and n values ('output' observations) of each variable which are put into a vector:

$$\mathbf{y} \quad : \quad [y_1 \quad y_2 \quad \cdots \quad y_n]^\mathsf{T}$$

$$\mathbf{x^{(1)}} \quad : \quad [x_1^{(1)} \quad x_2^{(1)} \quad \cdots \quad x_n^{(1)}]^\mathsf{T}$$

$$\mathbf{x^{(2)}} \quad : \quad [x_1^{(2)} \quad x_2^{(2)} \quad \cdots \quad x_n^{(2)}]^\mathsf{T}$$

$$\vdots \quad : \quad \vdots \quad : \quad \vdots \quad : \quad \vdots$$

$$\mathbf{x^{(p)}} \quad : \quad [x_1^{(p)} \quad x_2^{(p)} \quad \cdots \quad x_n^{(p)}]^\mathsf{T}.$$

As above with $p = 1$, we could assume a relationship between y and the independent variables $x^{(1)}, x^{(2)}, \ldots, x^{(p)}$ of the form

$$Y = \alpha + \beta_1 x^{(1)} + \beta_2 x^{(2)} + \cdots + \beta_p x^{(p)}.$$

Consider the case of $p = 3$ (or 3 'input' factors). The problem is to find

$$Y = \alpha + \beta_1 x^{(1)} + \beta_2 x^{(2)} + \beta_3 x^{(3)}$$

(that is, a particular set of values for coefficient vector $\mathbf{c}$) that minimises

$$R^2 = \|\mathbf{y} - \mathbf{Y}\|_2^2 = \sum_{i=1}^n [y_i - Y_i]^2,$$

where $\mathbf{c} = (\alpha \quad \beta_1 \quad \beta_2 \quad \beta_3)^\top$,

$$\mathbf{y} = \begin{pmatrix} y_1 \\ y_2 \\ \vdots \\ y_n \end{pmatrix}, \quad \mathbf{Y} = \begin{pmatrix} Y_1 \\ Y_2 \\ \vdots \\ Y_n \end{pmatrix}, \quad A = \begin{pmatrix} 1 & x_1^{(1)} & x_1^{(2)} & x_1^{(3)} \\ 1 & x_2^{(1)} & x_2^{(2)} & x_2^{(3)} \\ \vdots & \vdots & \vdots & \vdots \\ 1 & x_n^{(1)} & x_n^{(2)} & x_n^{(3)} \end{pmatrix}.$$

This least squares problem can be written in an equivalent form (suitable for MATLAB)

$$A\mathbf{c} = \mathbf{y}.$$

You must have realised that this is the \ method! Yes, and the same idea can be applied similarly to the higher order polynomial fitting (or other nonlinear function fitting) for this multiple variable case! As previously remarked, the `polyfit` method can no longer help. As it stands, `polyfit` only deals with the single variable case. In the project of Chapter 17, `polyfit2.m` will be developed for treating the $p = 2$ case.

To find the coefficient column vector $\mathbf{c} = [\alpha \quad \beta_1 \quad \beta_2 \quad \beta_3]^\top$ and the residual $R2 = R^2$, the MATLAB commands will be (using $\mathbf{x1}$ for the vector $\mathbf{x}^{(1)}$, $\mathbf{x2}$ for $\mathbf{x}^{(2)}$ and so on)

```
>>   A=[ones(size(x1)) x1 x2 x3];
>>   c=A\y;      Y=A*c;      r=y-Y;
>>   R2 = r'*r
```

5.2.3.1 Example data: `diabetic.m`

The M-file `diabetic.m` contains the data and a brief explanation of it so type

```
>>   help diabetic
>>   diabetic
```

The data sets will be kept as $\mathbf{y}$, $\mathbf{x1}$, $\mathbf{x2}$, $\mathbf{x3}$. You should find that the coefficients are $\mathbf{c}^\top = [36.9601 \quad -0.1137 \quad -0.2280 \quad 1.9577]$ and $R2 = R^2 = 567.6629$.

Then one may infer that the following function (that is, our least squares solution) describes the response to treatment

$$Y = Y(x^{(1)}, x^{(2)}, x^{(3)}) = 36.9601 - 0.1137x^{(1)} - 0.2280x^{(2)} + 1.9577x^{(3)},$$

where $x^{(1)} = $ age, $x^{(2)} = $ weight and $x^{(3)} = $ diet.

Exercises

5.1   The M-file `marks.m` contains the marks obtained by a class of engineering students, on a differential equations course, in a class test and in the homework prior to test. If you run the M-file by typing `marks`, you will generate a copy of the marks in matrix **mkdata**. Typing

```
>>   x=mkdata(:,1);
>>   y=mkdata(:,2);
```

pulls out the first and second columns of data and names them as **x** (test marks) and **y** (homework) respectively. Now try the following:

(a) Compute the least squares line (see §5.2)

$$Y = \alpha + \beta x$$

and the least squares line

$$X = \gamma + \delta y.$$

In this problem it is not possible to state that one variable is dependent and the other independent, but it is reasonable to assume some relationship. If the relationship was exact (that is, if the residual of the least squares fit was zero), then the two lines would be coincident. Find the equations of the two lines and describe briefly the method you used to obtain your answer.

(b) Plot the data as a scatter plot, and superimpose on it a plot of the two lines. Calculate the value of the angle between the lines.† Again, describe the method you used to obtain your answer.

Hint: you may use this formula

$$\tan(\theta_1 - \theta_2) = \frac{\tan\theta_1 - \tan\theta_2}{1 + \tan\theta_1 \tan\theta_2}.$$

5.2   The M-file `mannheim.m` gives some production figures for car power steering units together with the cost of production. After running `mannheim`, the data will be in a matrix called **mdata** of size 22×3. We shall name the first column of **mdata** as **x** and second column as **y**. Now try the following:

---

† The relationship between data sets is known as *correlation*. Here the angle is an indicator of how the data sets are correlated.

(a) Begin by sorting the data according to increasing **x**, naming the sorted data as **sx, sy**.

(b) Assume that y is the cost of production and x is the number of units produced. Fit a linear relation $y = \alpha + \beta x$ to the given data set.

(c) Define two new independent variables $x^{(1)}$ and $x^{(2)}$ as follows; we omit interpretation of their practical meanings

$$x^{(1)} = \begin{cases} x, & x < \gamma \\ \gamma, & x \geq \gamma, \end{cases} \qquad x^{(2)} = \begin{cases} 0, & x < \gamma \\ x - \gamma, & x \geq \gamma \end{cases}$$

with $\gamma = 3000$.

Firstly, work out the new data sets $\mathbf{x1} = \mathbf{x^{(1)}}$, $\mathbf{x2} = \mathbf{x^{(2)}}$. Secondly, use the method of §5.2.3 with the sorted data and $p = 2$ to compute a least squares fit of the form

$$y = \alpha + \beta_1 x^{(1)} + \beta_2 x^{(2)}$$

and find the residual of this fit. Finally, plot the nonlinear fit using `plot(sx,sy,'*',sx,Y)`, where **Y** is, as usual, the points of the nonlinear fit, as in §5.2.3. Does the graph appear to fit the data substantially better than the linear fit? Suggest a possible (real life) reason why the fit using two straight lines is so much better than the fit using one.

Hint What might happen once the number of units being produced exceeds a certain number? Try also `help mannheim`. Note that $x^{(1)} = \min(\gamma, x)$ and $x = x^{(1)} + x^{(2)}$ for all values of x.

EXTRAS Consider a further fitting

$$y = \alpha + \beta_{11} x^{(1)} + \beta_{21} x^{(2)} + \beta_{12}(x^{(1)})^2 + \beta_{22}(x^{(2)})^2$$

or alternatively

$$y = \alpha + \beta_1 e^{x^{(1)}} + \beta_2 e^{x^{(2)}}.$$

5.3 Appendix

Here we consider the minimisation of R^2 by selecting α and β; see §5.2.1. To proceed, we first prove the famous Cauchy–Schwarz inequality.

Theorem 5.1 (Cauchy–Schwarz inequality) *For any* $\mathbf{x}, \mathbf{y} \in \mathbf{R}^n$ *we have*

$$\sum_{j=1}^{n} |x_j y_j| \le \|\mathbf{x}\|_2 \|\mathbf{y}\|_2. \tag{5.1}$$

Proof (i) We only need to consider $\mathbf{x} \neq \mathbf{0}$ and $\mathbf{y} \neq \mathbf{0}$, as otherwise the result is trivially $0 = 0$;

(ii) We claim that it is sufficient to prove:

$$\sum_{j=1}^{n} x_j v_j \le \|\mathbf{x}\|_2 \|\mathbf{v}\|_2 \tag{5.2}$$

for any nonzero vectors $\mathbf{x}, \mathbf{v} \in \mathbf{R}^n$. For if this is true, we may construct $\mathbf{v}$ by $v_j = \text{sign}(x_j)\text{sign}(y_j)y_j$ for any $\mathbf{y} \in \mathbf{R}^n$ and $j = 1, \dots, n$ so that (5.1) is valid;

(iii) To prove (5.2), note that for *any* $\lambda \in \mathbf{R}$

$$0 \le \|\mathbf{x} + \lambda \mathbf{v}\|_2^2 = (\mathbf{x} + \lambda \mathbf{v})^\top (\mathbf{x} + \lambda \mathbf{v}),$$

so

$$2\lambda \sum_{j=1}^{n} |x_j v_j| \le \|\mathbf{x}\|_2 + \lambda^2 \|\mathbf{v}\|_2.$$

Now *select* a specific $\lambda = \|\mathbf{x}\|_2 / \|\mathbf{v}\|_2$, simplifying the above inequality as

$$2 \sum_{j=1}^{n} |x_j v_j| \le 2 \|\mathbf{x}\|_2 \|\mathbf{v}\|_2,$$

which proves inequality (5.2). Thus inequality (5.1) is proved. $\qquad\square$

5.3.1 Derivation of the least squares method

As R^2 is quadratic in α, β, the minimum values may be found by solving

$$\frac{\partial}{\partial \alpha} R^2 = 0 \quad \text{and} \quad \frac{\partial}{\partial \beta} R^2 = 0.$$

The least squares equations can be written out as

$$\begin{cases} \dfrac{\partial}{\partial \alpha} \sum_i (y_i - \alpha - \beta x_i)^2 = (-2) \sum_i (y_i - \alpha - \beta x_i) = 0, \\[2mm] \dfrac{\partial}{\partial \beta} \sum_i (y_i - \alpha - \beta x_i)^2 = (-2) \sum_i x_i(y_i - \alpha - \beta x_i) = 0, \end{cases}$$

that is,

$$
\begin{cases}
n\alpha + \left(\sum_i x_i \right) \beta = \sum_i y_i, \\
\left(\sum_i x_i \right) \alpha + \left(\sum_i x_i^2 \right) \beta = \sum_i x_i y_i.
\end{cases}
$$

Using the usual matrix notation, $\mathbf{c}^\top = (\alpha \ \beta)$,

$$
A = \begin{pmatrix} 1 & x_1 \\ 1 & x_2 \\ \vdots & \vdots \\ 1 & x_n \end{pmatrix}, \quad
\mathbf{x} = \begin{pmatrix} x_1 \\ x_2 \\ \vdots \\ x_n \end{pmatrix}, \quad
\mathbf{y} = \begin{pmatrix} y_1 \\ y_2 \\ \vdots \\ y_n \end{pmatrix},
$$

we can write (and define)

$$
\Phi(\mathbf{c}) = R^2(\mathbf{c}) = (\mathbf{y} - A\mathbf{c})^\top (\mathbf{y} - A\mathbf{c}).
$$

Then the above least squares equations reduce to

$$
A^\top A \mathbf{c} = A^\top \mathbf{y}. \tag{5.3}
$$

Mathematically speaking, to solve this 2×2 linear system, we first need to ensure

$$
\det(A^\top A) = n \sum_i x_i^2 - \left(\sum_i x_i \right)^2 \neq 0,
$$

which is true if at least two x_is are distinct! Then we can easily find the unique solution for $\mathbf{c} = (\alpha \ \beta)^\top$

$$
\begin{cases}
\beta = \dfrac{\displaystyle\sum_{i=1}^{n}(x_i - \bar{x})(y_i - \bar{y})}{\displaystyle\sum_{i=1}^{n}(x_i - \bar{x})^2}, \\
\alpha = \bar{y} - \beta \bar{x},
\end{cases}
$$

where $\bar{x}$ and $\bar{y}$ are the means of data sets $\mathbf{x}$ and $\mathbf{y}$ respectively.

To convince ourselves that the above solution is a minimum instead of a saddle point, we need to work out all the second derivatives (which are constants in this quadratic case). We may verify that

$$
\begin{cases}
\dfrac{\partial^2 \Phi}{\partial \alpha^2} = 2n, \qquad \dfrac{\partial^2 \Phi}{\partial \alpha \partial \beta} = \dfrac{\partial^2 \Phi}{\partial \beta \partial \alpha} = 2 \sum_i x_i, \\
\dfrac{\partial^2 \Phi}{\partial \beta^2} = 2 \sum_i x_i^2.
\end{cases}
$$

Then from the Cauchy–Schwartz inequality, the discriminant is negative since

$$
\Delta = \left(\frac{\partial^2 \Phi}{\partial \beta \partial \alpha} \right)^2 - \left(\frac{\partial^2 \Phi}{\partial \alpha^2} \right) \left(\frac{\partial^2 \Phi}{\partial \beta^2} \right) = 4 \left[\left(\sum_i x_i \right)^2 - n \sum_i x_i^2 \right] < 0,
$$

whenever $\det(A^\top A) \neq 0$, and both $\frac{\partial^2 \Phi}{\partial \alpha^2} > 0$ and $\frac{\partial^2 \Phi}{\partial \beta^2} > 0$ as at least two x_is are distinct (that is, they have no chance to be zero at the same time). This proves that the extreme point is indeed a minimum.

An alternative and simpler proof is as follows. Consider the following functional for all $\mathbf{w} = \mathbf{c} + \mathbf{h} \in \mathbf{R}^2$ (space)

$$
\Phi(\mathbf{w}) = (\mathbf{y} - A\mathbf{w})^\top (\mathbf{y} - A\mathbf{w}) = \mathbf{w}^\top A^\top A \mathbf{w} - 2\mathbf{y}^\top A\mathbf{w} + \mathbf{y}^\top \mathbf{y}.
$$

We can first verify that

$$
\Phi(\mathbf{c} + \mathbf{h}) = \Phi(\mathbf{c}) + (A\mathbf{h})^\top A\mathbf{h} + 2\mathbf{h}^\top (A^\top A\mathbf{c} - A^\top \mathbf{y}).
$$

Assuming $A^\top A\mathbf{c} = A^\top \mathbf{y}$, we then have

$$
R^2(\mathbf{c} + \mathbf{h}) = \Phi(\mathbf{c} + \mathbf{h}) = \Phi(\mathbf{c}) + \|A\mathbf{h}\|_2^2 \geq \Phi(\mathbf{c}) = R^2(\mathbf{c}).
$$

Therefore the solution to equation (5.3) is the unique minimum of the least squares problem.

6

Probability and Random Numbers

The study of probability and random, or stochastic, processes remains an important subject, simply because most real world problems exhibit random variations. Such variations can give rise to nondeterministic factors that definitive mathematics, even with approximations, cannot immediately describe. The *probability* of an event in classical probability theory is defined as the ratio of the frequency of the event to the total number of all possible outcomes. Often it is more convenient to model probability by a *density distribution* for both discrete and continuous random systems. Fortunately for many practical problems, suitable probability density functions, representing probability distributions, are known.

In this chapter, we introduce the elementary probability distributions and simulate them by generating appropriate random numbers with MATLAB.

6.1 Generating random numbers

The MATLAB function **rand** generates pseudorandom numbers on the interval $(0, 1)$. These numbers are *pseudorandom* because they *appear* to be random sequences but there is a method to duplicate them! The sequences are generated by a deterministic algorithm but they can be 'seeded' to yield a particular sequence.†

To see how this command works, try the following

```
>>  rand              % Generates a random number in (0,1)
>>  rand('seed',13) % Set the 'seed' of algorithm to 13
```

† The algorithm is based on a multiplicative congruential method. See the appendix to this chapter for details.

```
>> b1 = rand(25,1) % A random column vector in (0,1)
>> rand('seed',0)  % Reset the 'seed' to DEFAULT.
>> A = rand(3,4)   % 3x4 random matrix in (0,1)
>> rand('seed',13) % Reset the 'seed' to 13
>> b2 = rand(25,1) % Duplicate b1 !
```

It is possible to 'increase' the randomness by starting the sequence of numbers at a 'random' term, for example, by linking the `seed` to the internal clock of the computer. The current clock time is

```
>> time=round(clock) % round can be fix, ceil or floor
```

and this vector represents `time=[Year Month Day Hour Min Sec]` so we could use

```
>> rand('seed',time(6)) %or rand('seed',time(5)*time(6))
```

to link to the nearest second, though this doesn't always seem to work as well as setting the seed yourself, preferably to an *odd* number. Numbers in the range 10000–20000 (or actually the default 0) are good choices.

The simple MATLAB function **rand** generates a column of n random numbers uniformly distributed in $(0, 1)$ by `x=rand(n,1)`, or a row by `x=rand(1,n)`. Similarly, `rand(n)` generates a $n \times n$ random matrix while `rand(m,n)` produces a $m \times n$ matrix with entries in $(0, 1)$.

To generate random numbers in any interval (a, b), use a linear transformation $(\mathbf{x} \to \mathbf{y})$, for example

```
>> x = rand(1,30);      % 30 random numbers in (0,1)
>> a = 12; b = 99;      % set up the interval
>> y = a + (b-a) * x;   % 30 random numbers in (a,b)
```

Notice that the elements of vector **y** are now within (a, b).

6.2 Random integers

MATLAB functions `ceil`, `fix`, `floor` and `round`, combining with `rand`, generate random *integers*. Here `ceil(x)` is the smallest integer $\geq x$, `floor(x)` is the largest integer $\leq x$ and `round(x)` is x rounded to the nearest integer. The command `fix` mixes `ceil` and `floor` depending on the sign of a number, because it rounds numbers *towards zero* to the nearest integer. For instance, test the followings out

```
ceil(1.2)=2,   ceil(3)=3,   ceil(1.9)=2,   ceil(1.5)=2,
 fix(1.9)=1,    fix(2)=2,   fix(-1.6)=-1,   fix(-2)=-2,
```

```
floor(1.2)=1, floor(3)=3, floor(1.9)=1,  floor(1.5)=1,
round(1.2)=1, round(3)=3, round(1.9)=2,round(-1.5)=-2.
```

We can get a column of n numbers taking integer values $0, 1, \ldots, k$ (that is, integer values in the closed interval $[0, k]$) with equal probability using

```
>>  x = rand(n,1);   floor((k+1)*x)
```

We can get a column of n numbers taking integer values $1, \ldots, k$ (that is, integer values in the closed interval $[1, k]$) with equal probability using

```
>>  ceil(k*x)
```

The use of **round** is a little different. For example, the following

```
>>  round(10*x)
```

generates n numbers which do not take the values $0, 1, \ldots, 10$ with equal probability. (Can you see why? Compare the x-values rounded to 0 with those rounded to 1, and consider also **round(9.99 x + 0.5)**.)

If we simply want to generate a random permutation of k integers $1, 2, \ldots, k$, use the command **randperm**; for example, we may permute the rows of a matrix in a random manner

```
>>  r = randperm(9)
>>  A = ceil( rand(9)*5 )    % Generate a matrix in (1,5)
>>  B = A(r,:)               % Permute rows via "r"
```

Here is a cunning way to produce a column vector of 0s and 1s, where the 0s occur with 'probability $\frac{1}{3}$'. That is, for a very large vector, one third of them are likely to be 0 and the rest 1. Type

```
>>  rand('seed',0); x=rand(100,1);   % Note x in (0,1)
>>  c=(x>1/3)
```

The second line here instructs the computer to print 0 if **x** 'fails' the test $x > \frac{1}{3}$ and 1 if **x** 'passes'. Assuming that **x** has a uniform distribution in $[0, 1]$ you should get 0s in approximately $\frac{1}{3}$ of the cases. You can test this by typing **sum(c)/100** and seeing whether the answer is close to $\frac{2}{3}$.

However, a more challenging method than **c=(x>1/3)** is the following

```
>>  rand('seed',0); x=rand(100,1);   % Note x in (0,1)
>>  c=round(x + 1/6)
```

Here the trick is that, by a shift of $\frac{1}{6}$, only numbers in **x** less than $\frac{1}{2} - \frac{1}{6} = \frac{1}{3}$ are rounded to zero! Can you draw a picture to illustrate this idea? In a similar way, you can experiment with any probability p. Can you suggest a formula for the shift?

Example Starting from `seed=121`, to generate two random integer matrices of size 2×6 with elements in $[1, 4]$, and $[0, 3]$ respectively, we do the following

```
>>  rand('seed',121);  x = rand(2,6)
>>  y = ceil( 4*x )        % Case (1) with [1,4]
>>  z =floor( 4*x )        % Case (2) with [0,3]
```

Here are the results:

y =	1	2	2	1	3	2
	3	3	4	2	1	1
z =	0	1	1	0	2	1
	2	2	3	1	0	0

6.3 Simulating uniform distributions

We have discussed that the MATLAB function **rand** simulates numbers from a uniform distribution U(0,1).† As mentioned, we can use random numbers to simulate uniform distributions in any interval when we combine U(0,1) with a linear transformation; for your convenience we have developed an M-file called `unirand.m`.

Example To simulate 5000 rolls of three dice, we try

```
>> rand('seed',19)              % Set 'seed' to 19
>> z=unirand(0.5,6.499,5000,3);% Random in (0.5, 6.499)
>> r = round(z);                % Random numbers in [1,6]
>> [x y]=hist(r,6)              % Count faces in x
```

6.4 Simulating normal distributions

To obtain a column of 1000 random numbers with normal distribution having mean 0 and standard deviation 1, type

```
>>  randn('seed',12);   x=randn(1000);
```

† See the appendix to this chapter for a description of uniform distributions.

where `seed` is similar to that used for `rand`, but is specific to `randn`! Using again a linear transformation, to obtain numbers with mean m and standard deviation s, type

```
>>  y = m + s*x;
```

For your convenience we have developed an M-file called `normrand.m`.

Example To simulate heights of people in two different regions, both with mean $m = 3$ and standard deviation $s = 5$, we take a sample of 2000 people from each region and do the following

```
>> randn('seed',11)           % Set 'seed' to 11
>> y = normrand(3,5,2000,2);% N(3,5) with mean=3 & Std=5
>> hist(y)                    % Plot the simulated data y
```

or without `normrand.m`

```
>>  randn('seed',11)       % Set 'seed' to 11
>>  x = randn(2000,2);     % N(0,1) of mean=0 & Std=1
>>  y = 3 + 5*x;           % Transformation
>>  hist(y)                % Plot the simulated data y
```

and you should find that $\text{mean}(\mathbf{x}) = [0.0318 \;\; -0.0346]$, $\text{mean}(\mathbf{y}) = [3.1591 \;\; 2.8270]$ and $\text{std}(\mathbf{y}) = [4.9989 \;\; 5.0467]$.

Again as with `rand`, to obtain random integers, combine `randn` with any of the functions `ceil`, `fix`, `floor` and `round`. To get, for example, seven columns of nine whole numbers with normal distribution of mean $m = 3$ and standard deviation $s = 4$, type

```
>>  n1 = 9;  n2 = 7;  x = randn(n1,n2); m=3; s=4;
>>  y = round(m + s*x);
```

Remark: The two kinds of distributions are distinguished by `rand` and `randn`. By default, `rand` produces a uniform distribution. However, commands `rand('normal')` and `rand('uniform')` can force `rand` to toggle between the two distributions! We discourage this practice, as one may lose track of this toggling unless this is set each time `rand` is used!

6.5 Simulating negative exponential distributions

As the negative exponential distribution is related to the uniform distribution,† its simulation can be done easily using the MATLAB random

---

† See the appendix to this chapter.

number generator `rand`. To simulate a negative exponential distribution, with both *mean* and *standard deviation* being ℓ, we can use the following relation

$$\mathbf{z} = -\ell \log(1 - \mathbf{x}),$$

assuming that $\mathbf{x}$ is from a uniform distribution U(0,1). For your convenience we have developed an M-file `exprand.m`, generating this distribution.

To illustrate, we can simulate and compare samples of $n = 20000$ numbers of

- a uniform distribution in $(0, 2.4)$ with mean $m = 1.2$;
- a normal distribution with mean $m = 1.2$ and standard deviation $s = m = 1.2$;
- an exponential distribution with mean $\ell = 1.2$:

```
>> rand('seed',0); n=20000;  % Reset the 'seed' value
>>  m=1.2; s=1.2;            % mean and std
>>  x = unirand(0,2.4,n,1);  % Uniform in (0,2.4)
>> randn('seed',0);          % Reset the 'seed' value
>>  y = normrand(m,s,n,1);   % Normal (m,s)=(1.2,1.2)
>> rand('seed',0);           % Reset the 'seed' value
>>  z = exprand(m, n,1);     % Exponential (mean m)
>>  subplot(131); hist(x,90) % Plot x
>>  subplot(132); hist(y,20) % Plot y
>>  subplot(133); hist(z,40) % Plot z
```

This example is available in the M-file `c6exp.m`. The results can be seen in Figure 6.1, where one graph shows an exponential distribution! The mean and standard deviation of a negative exponential distribution are both $\ell = m$ and so you may use MATLAB commands `mean` and `std` to verify that, for $\mathbf{z}$, both quantities are close.

As discussed in §6.6, in practice, inter-arrival times satisfy the negative exponential distribution. For instance, take the example of five arrival times (occurrences) of an event as `arr` $= [1.3\ 2.4\ 5.1\ 6\ 8.3]$. Then inter-arrival times are the time differences between arrival times (occurrences), that is, `int_arr` $= [1.3\ 1.1\ 2.5\ 0.9\ 2.3]$ counting from $t = 0$.

So we can simulate inter-arrival times by `exprand.m` but how do we work out arrival times then? Each arrival time is a cumulative sum of inter-arrival times; for the above example, given `int_arr`, we work out `arr` by

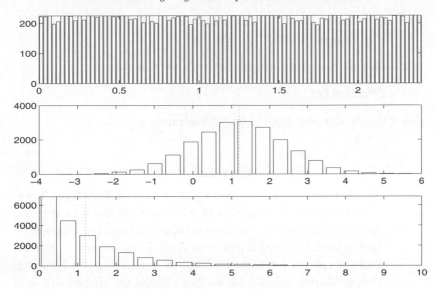

Fig. 6.1. Illustration of three probability distributions.

```
arr(1) = int_arr(1) = 1.3;
arr(2) = int_arr(1) + int_arr(2) = 2.4;
arr(3) = int_arr(1) + int_arr(2) + int_arr(3) = 5.1;
arr(4) = sum of 'int_arr(1) +...+ int_arr(4)' = 6;
arr(5) = sum of 'int_arr(1) +...+ int_arr(5)' = 8.3.
```

In MATLAB, the command for cumulative sums is simply

```
>> t = cumsum(z) ;
```

For the above example in Figure 6.1, the cumulative sum for z can be written as

$$t_k = \sum_{i=1}^{k} z_i, \qquad k = 1, 2, \cdots, 20000.$$

Here $t(1) = 0.2966$ and $t(20000) = 24036.2043$.

Example There is a concern about the long queues at a bank cash machine and we have been asked to simulate the arrivals during peak times of the day. It is known from computer records that (on average) there are 131 customers between 9 am and noon. In order to simulate the arrival of the first 50 customers, first work out the mean $\ell = \frac{180}{131}$ since the average inter-arrival time is $\frac{180}{131}$ minutes. Hence type

```
>>  m = 180 / 100;        rand('seed',0);
>>  z = exprand(m,50,1);   hist(z)
```

As z is the *inter-arrival time*, the arrival time t is simply

```
>>  t=cumsum(z)
```

with $t(1) = 0.3396$ and $t(50) = 69.3055$ minutes.

Exercises

6.1   An electrical retailer estimates that, in the weeks before Christmas, there are, on average, 140 customers per day for electronic games, with inter-arrival times satisfying a negative exponential distribution. (A shop day is 9 am–5 pm.) On average $\frac{2}{3}$ pay cash and the rest use a credit card, 50% of the sales are for Nintendo (Super Mario) and 50% are for Sega (Sonic the Hedgehog). You are asked to simulate (the first) 30 sales from the opening time, assuming that each customer buys a product.

(a) Simulate the inter-arrival and arrival times for the first 30 customers of the day. Find the mean and standard deviation for the inter-arrival times.

(b) Print out a table simulating the first 30 sales in the form of *Arrival Time*, in minutes with 1 decimal place, *Payment Method* (Cash as +1 and Credit as −1) and *Product Sold* (Nintendo as +1 and Sega as −1)
Hints:

- Your answer to question (b) will be three columns (all of size 30):
  one set of arrival times,
  one set to select cash or credit card and
  one set to select Nintendo or Sega.
- Consult §6.2 to see how to obtain numbers which are 0, 1 with probability $\frac{1}{3}, \frac{2}{3}$, or both with probability $\frac{1}{2}$. How do you then convert such a vector of 0s and 1s into −1s and +1s ?
- MATLAB cannot print different columns of the same matrix in different formats; they all need to be integers or all decimals. To change this you need to use the command `fprintf` or `sprintf`. For example, try

```
>>   a = pi,   b = 35,   c = 15
>>   c_1 = [ a b c ]
>>   fprintf('c_2 = %5.3f   %5d   %d\n',a,b,c)
>>   c_3 = sprintf('%5.3f   %5d   %d\n',a,b,c)
```

where '%5.3f' means printing a real number with three decimals in a total length of five, '%5d' means printing an integer with five digits while '%d' prints a default number of digits. To print the *Arrival Time*, you need '%6.1f'.

6.2   It is estimated that, on average, 15 trains per hour arrive at Liverpool Lime Street station, with inter-arrival times satisfying a negative exponential distribution. In addition it is known that 75% of the trains are small local diesel units on which the number of passengers (per train) is uniformly distributed on $[0, 80]$. The remaining 25% are inter-city trains on which the number of passengers is assumed normally distributed with mean 90 and standard deviation 20.

(a) Make a 15×5 matrix simulating the train types ($0 =$ inter-city, $1 =$ diesel) for the first 15 trains on 5 days. (See §6.2 again for obtaining a column of numbers which take the value 0 with probability $\frac{1}{4}$ and 1 with probability $\frac{3}{4}$.)

(b) Use the given distributions to simulate the numbers of passengers on 15 diesel trains and 15 inter-city trains for 5 days. (So your answer will be two matrices, each 15×5, one matrix for each train type.)

(c) Combine (a) and (b) to simulate the total number of passengers on the first 15 trains arriving at Lime Street, for 5 days.

(d) Simulate the arrival times of the first 15 trains on 5 days, and use this information and (c) to estimate the total number of passengers arriving in the first half-hour of each day.

(e) Make a plot of the cumulative passenger totals against arrival times for the five days on the same graph.
Hint If the arrival times are in a 15×5 matrix **arr** and the cumulative passenger totals are in a 15×5 matrix **cumpass** then **plot(arr,cumpass)** plots all five graphs at

once. From `help plot` or `arr(15,:)` and `cumpass(15,:)`, can you distinguish the curves?

6.6 Appendix

6.6.1 Generation of random numbers

The popular approach of generating random numbers is by the so-called *multiplicative congruential method*. It uses the equation

$$u_{i+1} = Ku_i(Mod \ M), \qquad\qquad i = 1, 2, 3, \ldots,$$

where '*Mod M*' means that we subtract as many multiples of M as possible, and take the *remainder* as our answer. For any i, u_i is a random number such that $0 \le u_i < M$. Therefore u_i corresponds to a random number $r_i = u_i/M \in [0, 1)$. To use this method, once suitable constants K, M (usually integers) are chosen, we only need to supply an initial value u_1 which is usually called a 'seed'! Usually K is of about the same magnitude as the square root of M but M, although large, must not be a multiple of K.

With MATLAB, the method for any one step is simply implemented as

```
>>  unew = rem ( K * uold, M )
>>  rand_num = unew / M
```

Taking $u_1 =$ `seed` which is supplied by the user, we can represent n steps of a *congruential method* as follows

```
>>  for j = 1:n
>>  seed = rem ( K * seed + shift,  M )
>>  rand_num = seed / M
>>  end
```

where `shift` is zero in the standard method but can be nonzero!

For the benefit of interested readers, we have developed an M-file `randme.m` for implementing this method. Type `randme` to see what it does! For example we have used `a=randme(4)` with `seed=2` and `shift=0`, taking $K = 32$ and $M = 5$, to produce the following 'home-made' four random numbers

```
K*seed = 64,    seed = 4,    rand = 0.8
K*seed =128,    seed = 3,    rand = 0.6
K*seed = 96,    seed = 1,    rand = 0.2
```

K*seed = 32,   seed = 2,   rand = 0.4

The random number generator `rand` adopted by MATLAB also uses the above method. To find out what the current `seed` is type `s = rand('seed')`, while `rand('seed',45)` is to change the `seed` to 45. Therefore, to observe a sequence of `seed` changes in MATLAB type

```
>>  rand('seed',0);        % Fix the seed to a known value
>>  s = rand('seed')       % Find out the default seed!
>>  rand('seed',1);        % Fix the seed to a known value
>>  s = rand('seed');      % Check that it is indeed there
>>  fprintf('MATLAB Seed = %12.0f (initially set)\n',s)
>>  for j = 1:5, r=rand; s=rand('seed');
>>  fprintf('MATLAB Seed = %12.0f  rand = %d',s,r);end
```

One MATLAB manual seems to suggest $K = 7^5$ and $M = 2^{31} - 1$ but, experimenting with `randme.m` and these two parameters, one may conclude that there must be an unpublished integer `shift` involved! For our purpose, we only need to know the basic algorithm

$$\text{seed} = (K\text{seed} + \text{shift}) \, (Mod \ M).$$

6.6.2 Distribution functions

The distribution of the pseudorandom numbers is continuous (up to machine precision) and uniform (equally likely in the range $(0, 1)$). For any distribution of random numbers, the *probability* of sampling a particular value is governed by the *probability density function* (pdf) f so that if we denote by X the random variable, then for any values x_1 and x_2 such that $x_1 < x_2$, it follows that

$$\mathcal{P}(x_1 \leq X \leq x_2) = \int_{x_1}^{x_2} f(t)dt.$$

Alternatively a distribution can be defined in terms of the *cumulative distribution function* (cdf) F, where

$$F(x) = \mathcal{P}(X \leq x);$$

the relation between the two functions is

$$F(x) = \int_{-\infty}^{x} f(t)dt \qquad \text{with} \qquad \int_{-\infty}^{\infty} f(t)dt = 1.$$

Uniform distributions For a uniform distribution on the interval $[0, 1]$, known as the normalised uniform distribution U(0,1), the pdf is

$$f(x) = \begin{cases} 1, & x \in [0, 1], \\ 0, & \text{otherwise,} \end{cases}$$

and the cdf

$$F(x) = \begin{cases} 0, & x < 0, \\ x, & x \in [0, 1), \\ 1, & x \geq 1. \end{cases}$$

Normal distributions The distribution of a normal random variable X is governed by the pdf

$$f(x) = \frac{1}{\sigma\sqrt{2\pi}} \exp\left\{ -\frac{(x-\mu)^2}{2\sigma^2} \right\}$$

or equivalently the cdf

$$F(x) = \frac{1}{\sigma\sqrt{2\pi}} \int_{-\infty}^{x} \exp\left\{ -\frac{(t-\mu)^2}{2\sigma^2} \right\} dt.$$

This distribution has the mean μ and standard deviation σ, denoted by $N(\mu, \sigma^2)$.

When $\mu = 0$ and $\sigma = 1$, N(0,1) is the so-called standard normal distribution which is simulated by **randn** in MATLAB. The distribution $N(\mu, \sigma^2)$ can be obtained by the transformation $z = \sigma x + \mu$, if x is the standard normal variable N(0,1).

Negative exponential distributions A negative exponential distribution describes a Poisson process. Recall that a series of events (denoted by random variable X) is a Poisson process if

- the number of outcomes in each period is independent;
- the probability of more than one event in such a small interval is negligible;
- the probability of a single event during a (very short) interval is proportional to the length of the interval, that is,

$$\lim_{h \to 0} \frac{\mathcal{P}\{\text{an event occurs in the time interval}(t, t+h)\}}{h} = \lambda,$$

where $\mathcal{P}$ denotes the probability and λ is fixed.

The distribution is governed by the pdf

$$f(z) = \begin{cases} \lambda e^{-\lambda z}, & z \geq 0, \\ 0, & \text{otherwise,} \end{cases}$$

or equivalently the cdf

$$F(z) = \begin{cases} \int_0^z e^{-\lambda t} dt = 1 - e^{-\lambda z}, & z \geq 0, \\ 0, & \text{otherwise.} \end{cases}$$

As the cdf is in a simple and closed form, we can link it with the uniform distribution (another simple cdf) and thus do not need a separate random number generator. If x is selected from a uniform random distribution, to generate an exponential distribution z, let

$$1 - e^{-\lambda z} = x$$

giving $(\ell = \lambda^{-1})$

$$z = -\ell \log(1 - x).$$

This is actually what M-file `exprand.m` does. For the bank example in §6.5, where there are 131 customers (arrivals) in 180 minutes, the average inter-arrival time is $\ell = h_o = \frac{180}{131}$ minutes. Therefore, the probability of an arrival (event) in any short interval $(t, t + h_o)$ is $\mathcal{P} = 1$ so

$$\lim_{h \to 0} \frac{\mathcal{P}\{(t, t+h)\}}{h} \approx \frac{\mathcal{P}\{(t, t+h_o)\}}{h_o} = 1/h_o = 131/180 = \lambda.$$

Of course, $\ell = 1/\lambda$ as expected.

7
Differential and Difference Equations

In this chapter we show how MATLAB can be used to explore, and to solve numerically, various simple types of differential equation.

7.1 Ordinary differential equations (ODEs)

You will probably have come across the differential equation

$$\frac{dx}{dt} = -\lambda x \tag{7.1}$$

which is easily solved by writing it as

$$\frac{dx}{x} = -\lambda dt$$

and integrating both sides. This gives

$$\ln x = -\lambda t + c$$

that is,

$$x(t) = x_0 e^{-\lambda t}, \qquad \text{where} \quad x_0 = x(0). \tag{7.2}$$

This describes, for example, radioactive decay (positive λ) or exponential growth (negative λ).

The 'dot' notation for differentiation is often used when the independent variable is t, as in this case. Thus equation (7.1) is often written, in compact notation as

$$\dot{x} = -\lambda x.$$

Another interesting example, but a little harder to solve, is the logistic equation

$$\dot{x} = rx \left(1 - \frac{x}{K} \right) \tag{7.3}$$

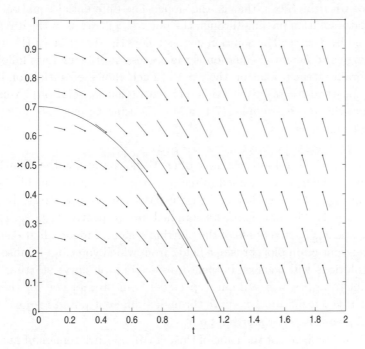

Fig. 7.1. Grain plot for the ODE $dx/dt = -x$. The curve is a particular numerical solution.

describing, for example, the rate of infection in a population. The solution is

$$x(t) = \frac{K}{1 - (1 - \frac{K}{x_0})e^{-rt}}.$$ (7.4)

MATLAB can be used to solve these and other more complicated ODEs by a numerical integration procedure. Its graphical capabilities can also be used to study visually the behaviour of solutions. An M-file `fodesol.m` is provided to demonstrate this. As usual, typing

```
>> help fodesol
```

gives some information on how it is used.

The first order differential equation $dx/dt = f(x, t)$ can be thought of as defining a 'grain' on the t, x plane like the grain of a piece of wood, or the magnetic lines due to a bar magnet. See Figure 7.1 for an example.

The direction of the grain at the point (t, x) is defined by $f(x, t)$, which, being equal to dx/dt, gives the tangent of the angle the grain makes with the t-axis, that is, the slope. The differential equation tells us that a solution passing through the point (t, x) travels in the direction defined by $f(x, t)$. The solution curves follow the direction of the grain. Solving the differential equation is the process of drawing lines following this grain at each point on their path. Each choice of initial condition ($x = x_0$ when $t = t_0$) gives a different path (that is, solution). You can explore the above examples (7.1) and (7.3) using `fodesol`.

To begin type

```
>> fodesol('fnxt',0,2.5,-3,3)
```

where `fnxt.m` is a simple example of an M-file containing the right-hand side of the ODE to be studied (already provided for you). In this case it supplies $f(x, t) = -x$, that is, the right-hand side required to study (7.1) when $\lambda = 1$. The four numbers supplied are, respectively, t_{min}, t_{max}, x_{min} and x_{max}, the limits of the plotting region. `fodesol` first displays the relevant grain plot, or 'slope field', from which you can visualise how the solutions will behave. It then draws a solution curve starting from a point (t_0, x_0) which you must specify. It will prompt you for these in turn. If you want another curve through a different point, respond with y and give it another starting point.

When you have got the hang of this, create another version of `fnxt.m`, say `myfnxt.m`, which contains the right-hand side of the logistic equation (7.3) where, say,

$$K = r = 1 \quad .$$

Of course you can use `fodesol` to study any first order equation irrespective of the original choice of variable name used for the independent and dependent variables (here t and x respectively).

Explore solutions $x(t)$ which have $x_0 = x(t_0) < 1$ and also those with $x_0 > 1$.

Another interesting ODE to look at is

$$\frac{dx}{dt} = \frac{x}{t} - x, \tag{7.5}$$

for example, for

$$-3 < t < 4, \quad -3 < x < 4 \quad .$$

You should create another version of `fnxt.m` and look at its solutions.

7.2 Systems of differential equations

MATLAB can also numerically solve systems of coupled first order differential equations. For example, the equations

$$dx/dt = f(x,y), \qquad dy/dt = g(x,y) \qquad (7.6)$$

are first order and coupled since the derivative of x, say, depends on y as well as on x. They are also a bit special since t is not explicitly involved on the right-hand side. They are called *autonomous* differential equations for this reason. Equations like these are widely used to describe the interaction of competing species as shown in the example below equations (7.7). First some theory.

7.2.1 The phase plane

Equations (7.6) describe the motion of a point with coordinates (x,y) in the x,y plane, called here the phase plane. See Figure 7.2 for an example. As time changes, the point (x,y) will move unless of course both $f(x,y) = 0$ and $g(x,y) = 0$. If this is the case then the point (x,y) will remain stationary. Such stationary points are called *fixed points* of the system of differential equations. Thus the fixed points can be found from solving $f(x,y) = 0$ and $g = 0$. If this is not easy to do, experimenting with different starting values for x and y may indicate where possible fixed points may lie.

Near a fixed point there are a number of possibilities:

(i) The path followed by the solution may always move closer to the fixed point. The point is then said to be *stable*.

(ii) The path may approach the point and then move away. The point is then called a *saddle point* and is unstable.

(iii) The path may always move away from the point. Such points are also called *unstable*.

(iv) The path may just move around the point in some sort of *orbit*.

You should now see if you can identify these behaviours in the following example.

7.2.2 Competing species

The M-files `species.m` models the behaviour of two different creatures which compete for the *same* food source. This is modelled by the equa-

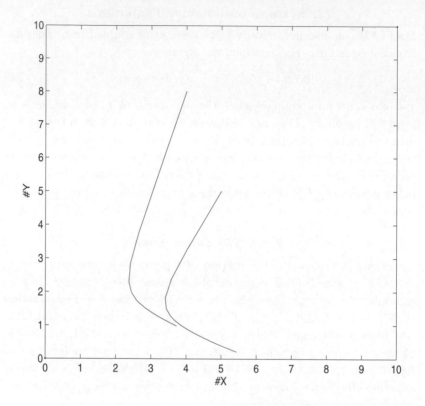

Fig. 7.2. The phase plane for the system of equations 7.7 for some choice of parameters. The curves are particular numerical solutions.

tions

$$\dot{x} = x(a - bx - cy), \qquad \dot{y} = y(d - ex - fy) \qquad (7.7)$$

where x and y are the numbers of the two different species.

The parameters a, b, c and d determine the behaviour of the system. For some values you will find that one species always wipes out the other. For different values the species can coexist, while for a further set the winning species depends on the initial populations, and so on.

The four types of fixed point behaviour described above are illustrated by the sets of parameters given below. Investigate each of these cases. To begin just type `species`.

(i) $a = 24$ $b = 6$ $c = 8$ $d = 24$ $e = 8$ $f = 6$
(ii) $a = 24$ $b = 4$ $c = 6$ $d = 24$ $e = 6$ $f = 8$

(iii) $a = 6$ $b = 0$ $c = 1$ $d = -4$ $e = -1$ $f = 0$

See if you can identify an example of each type of fixed point. An important first step is to identify the fixed points that is, where, according to equation (7.7),

$$\dot{x} = \dot{y} = 0.$$

This means you need to solve

$$
\begin{aligned}
x(a - bx - cy) &= 0, \\
y(d - ex - fy) &= 0
\end{aligned}
$$

for the fixed points (x, y) in terms of $a, b, c, \ldots, f$. You then just need to substitute values for a, b etc to get the fixed points (at most 4 of these) for each case. This helps you start your exploration. In each case, how many different solutions (fixed points) did you find which had $x \geq 0$ and $y \geq 0$? You are, of course, only interested in solutions with non-negative numbers of each species! The differential equations may well have solutions which are not of relevance to the physical application at hand.

Use `species` to explore the neighbourhood of *each* of the fixed points noting the behaviour close to them. The stable points are easy to find but the unstable points are more difficult. Notice that as you get nearer to a fixed point the motion of the point slows down and when you go further away it speeds up. As well as a graph, the M-file displays the coordinates at the end of each time-step.

7.2.3 Higher order equations

High order equations can be written as a system of first order ODEs, by a simple trick using derivatives. Then the MATLAB/numerical ODE solver can be used to give numerical solutions. Consider the following second order differential equation in t

$$\frac{d^2 w}{dt^2} - k(1 - w^2)\frac{dw}{dt} + w = 0, \tag{7.8}$$

called the Van der Pol equation (k is some positive constant). It describes oscillations in some electrical component. It can be converted into an autonomous system with variables v and w by setting $v = dw/dt$. Thus you get the pair of first order equations

$$\frac{dw}{dt} = v,$$

$$\frac{dv}{dt} = k(1 - w^2)v - w\,. \tag{7.9}$$

The M-file `vderpol.m` illustrates the solution of this equation. Run it for different values of the parameter k (just type `vderpol` and follow the instructions). For very small k, say about 0.1, you will find a *limit cycle* which is almost circular. In a limit cycle: the solution path settles down into a closed loop of some shape, irrespective of its initial starting point. Note that this is not the same as the *orbit* behaviour described in §7.2.1, where the solution curve always closes on itself. For larger values of k the shape of the limit cycle is very different.

Have a look at the M-files `vderpol.m` and `vdplfn.m`. The latter provides the right-hand side of the coupled equations (7.9). You can adapt these to solve your own higher order differential equations as needed, for example the equation of a damped harmonic oscillator or a projectile in a resistive medium. One of the projects in Part three of the book (Chapter 22) involves doing this amongst other things. When making modifications of the files `vderpol.m` and `vdplfn.m` for solving your own systems of equations make your own copies and rename them: for example `mysolver.m` and `myslvfn.m`. This should help to avoid confusion with the original M-files and to ensure that you are solving the equation you mean to solve!

7.3 Difference equations

The radioactive decay equation (7.1) expresses the fact that, in unit time, a constant fraction λ of the number of atoms present (x) is likely to decay. Calculus allows us to make this statement precise over an arbitrarily small time interval. In many applications, it is convenient to stick to a particular time interval and discuss *differences* $x(t+1) - x(t)$ rather than derivatives which describe instantaneous rates of change. In the case of radioactive decay we could write an (almost) equivalent *difference equation*

$$x_{n+1} - x_n = -\lambda x_n\,. \tag{7.10}$$

Here, n is a positive integer labelling the time measured in units of, say, 1 second, that is,

$$x(t) = x_n, \quad t = n \times 1\,\text{s}\,.$$

It is not hard to verify that

$$x_n = x_0(1 - \lambda)^n \tag{7.11}$$

is a solution of the difference equation (7.10) since the latter can be written as

$$\frac{x_{n+1}}{x_n} = 1 - \lambda.$$

For *small* values of λ you can also check that this solution is equivalent to the differential equation solution (7.2) since, in this case,

$$e^{-n\lambda} \approx 1 - n\lambda \approx (1 - \lambda)^n.$$

Difference equations may, of course, be studied as mathematical tools in their own right. In many applications, they arise more naturally out of the problem than do differential equations. In situations involving discrete variables, for example, one may wish to solve the difference equation directly rather than make some continuous approximation.

The classification and nomenclature of difference equations are similar to those used for differential equations. In particular, the treatment of inhomogeneous difference equations is the same. One adds a particular solution ('particular integral') to the solution of the homogeneous equation.

Here is a typical example of a homogeneous, linear difference equation that can be solved easily by a trial method similar to that used to solve the corresponding constant coefficient ODE:

$$x_{n+2} - 5x_{n+1} + 6x_n = 0. \tag{7.12}$$

This is a second order equation since *two* differences $x_{n+2} - x_{n+1}$ and $x_{n+1} - x_n$ are involved. (Try rewriting (7.12) in terms of differences.) We look for a solution similar to (7.11)

$$x_n = z^n. \tag{7.13}$$

Substituting we get

$$z^{n+2} - 5z^{n+1} + 6z^n = 0$$

or

$$z^n(z^2 - 5z + 6) = 0.$$

Ignoring the trivial solution $z = 0$, we solve the quadratic factor to find

$$z = 2 \quad \text{or} \quad z = 3.$$

So we get two solutions of the form (7.13) which must be added in the usual way to give

$$x_n = A2^n + B3^n$$

as the general solution. Notice there are *two* arbitrary constants since this is a *second* order difference equation. Just as for differential equations, analytic solutions like this one are not always easy to obtain. However, it is usually simple to get the computer (for example using MATLAB) to perform the iterations implied by such equations directly. The M-file `diffeqn.m` is set up to solve the above example in this way. To begin type

```
>> diffeqn
```

and follow the instructions. You should look inside this M-file and `dfeqfn.m` so you can understand how to use or modify them to solve other similar homogeneous equations, or indeed general inhomogeneous equations.

Try various different initial conditions including $[x_0, x_1] = [1.0, 2.0]$, $[1.0, 1.8]$ etc.

Exercises

7.1   Use `fodesol.m`, and a modified version of `myfunxt.m`, to study solutions of the following equation

$$\frac{dx}{dt} = x(1 - x^2). \tag{7.14}$$

In particular:

(a) Find $y(2)$ given the initial condition $y(0) = 1.5$.
(b) Find $y(3)$ given the initial condition $y(0) = 0.055$.
(c) Sketch the general behaviour of all the different types of solutions which you expect, having studied the 'slope field' or 'grain plot'.

7.2   From the competing species examples in §7.2.2, identify an example of each of

(a) fixed point surrounded by an *orbit*;
(b) *stable* fixed point;
(c) *saddle* point.

In each case make a *rough* sketch of the behaviour of the trajectories (solutions) in the neighbourhood of the fixed point, that is, showing the *direction* of flow of the solutions.

7.3   Modify the M-files `vderpol.m` and `vdplfn.m` to study the (non-autonomous) second order differential equation

$$\frac{d^2w}{dt^2} - a(w^2 - t) = 0,$$

where a is some parameter. (This is 'non-autonomous' since it has a term which is an explicit function of t, but it can be solved in the same way.) For the case $a = 3$, find the value of $w(4)$, and give a *rough* sketch of $w(t)$, for each of the solutions satisfying:

(a) $w(0) = 0$, $\dot{w}(0) = 1$;
(b) $w(0) = -1$, $\dot{w}(0) = 1$;
(c) $w(0) = 0$, $\dot{w}(0) = -1$.

Suggest initial conditions which give a solution $w(t)$ which is monotonically increasing.

7.4   Use `diffeqn.m` to solve the difference equation

$$6x_{n+2} - x_{n+1} + 2x_n = n^2 + 3n + 2$$

subject to $x_0 = 1$ and $x_1 = 2$ and, in particular, find x_{10}. Verify your solution by comparing the left- and right-hand sides of the equation for $n - 8$.

Part two
Investigations

8

Magic Squares

Aims of the project

Magic squares have been known for centuries. This project explores their properties from the perspective of matrix algebra, that is, using addition and multiplication of matrices. The project is not concerned with the number-theoretic problem of finding magic squares containing consecutive integers. The project is self-contained, but it may be of interest to know that several of the mathematical results come from the article [16]. This article also contains other results on the same subject.

Mathematical ideas used

Matrix multiplication, row reduced echelon form and solution of linear equations are used. Also, for example, 3×3 matrices are regarded as lying in nine-dimensional space $\mathbf{R}^9$, and subspaces of $\mathbf{R}^9$ are considered. (There is no requirement to know the definition of an abstract vector space: all spaces are contained in some $\mathbf{R}^n$.) The ideas of linear independence and basis are used. It is necessary to know that, in a subspace X of dimension r in $\mathbf{R}^n$, a set of r vectors in X which is linearly independent automatically spans X and so forms a basis. It is necessary to know the definitions of eigenvalues and eigenvectors, and to use these in a simple argument involving powers of matrices.

MATLAB techniques used

The project is about matrices, so you will need the techniques described in Chapter 2. At one point there is an M-file with several 'for' loops, and 'if' statements, so you will need to understand these ideas. See Chapter 3. Note that the project is somewhat 'open-ended': students who work quickly might like to go on to the final section, on 5×5 magic squares, which could be regarded as optional.

8.1 Introduction

This project is about magic squares. An $n \times n$ magic square is an $n \times n$ matrix of real numbers with the following property:

All rows, all columns and the two 'main' diagonals of the matrix add up to the same number, r say, called the magic constant.

For example,

$$\begin{pmatrix} 3 & 1 & 2 \\ 1 & 2 & 3 \\ 2 & 3 & 1 \end{pmatrix}$$

has this property for $r = 6$. The two main diagonals in this case are $3 + 2 + 1$ (top left to bottom right) and $2 + 2 + 2$ (top right to bottom left).

There are many algorithms for producing magic squares with the additional property that the entries are the integers $1, 2, \ldots, n^2$ in some order. For example, with $n = 3$,

$$\begin{pmatrix} 8 & 1 & 6 \\ 3 & 5 & 7 \\ 4 & 9 & 2 \end{pmatrix}. \tag{8.1}$$

The MATLAB function `magic` does this (try typing `magic(3)`). In this project we shall not go into these algorithms but instead investigate the algebra underlying magic squares, using your knowledge of matrices and solution of linear equations.

8.2 Magic squares size 3 × 3

(i) Consider a general 3×3 matrix

$$A = (a_{ij}) = \begin{pmatrix} a_{11} & a_{12} & a_{13} \\ a_{21} & a_{22} & a_{23} \\ a_{31} & a_{32} & a_{33} \end{pmatrix}.$$

Thus there are nine entries $a_{11}, \ldots, a_{33}$. Explain why, writing down the conditions for all rows, all columns and the two main diagonals to add to the same number r, we get the condition $M\mathbf{v} = 0$, where M is the

8 × 10 matrix

$$\begin{pmatrix} -1 & 1 & 1 & 1 & 0 & 0 & 0 & 0 & 0 & 0 \\ -1 & 0 & 0 & 0 & 1 & 1 & 1 & 0 & 0 & 0 \\ -1 & 0 & 0 & 0 & 0 & 0 & 0 & 1 & 1 & 1 \\ -1 & 1 & 0 & 0 & 1 & 0 & 0 & 1 & 0 & 0 \\ -1 & 0 & 1 & 0 & 0 & 1 & 0 & 0 & 1 & 0 \\ -1 & 0 & 0 & 1 & 0 & 0 & 1 & 0 & 0 & 1 \\ -1 & 1 & 0 & 0 & 0 & 1 & 0 & 0 & 0 & 1 \\ -1 & 0 & 0 & 1 & 0 & 1 & 0 & 1 & 0 & 0 \end{pmatrix}$$

and **v** is the column vector

$$(r \ \ a_{11} \ \ a_{12} \ \ a_{13} \ \ a_{21} \ \ a_{22} \ \ a_{23} \ \ a_{31} \ \ a_{32} \ \ a_{33})^{\mathsf{T}}.$$

(This really *is* just a matter of blindly applying the definition!)

(ii) Use the MATLAB function `rref` to find the reduced row echelon form of the matrix M. Why does it follow that, if A is to be magic, then *three* of the entries in the matrix A, say those in the bottom row, can be chosen arbitrarily and the rest are then determined? (Note that, after row reduction of M, the top row of the reduced matrix can be ignored since it just gives the equation

$$r = \text{ sum of elements in bottom row of } A,$$

and this is now the only equation involving r.)

Another way of saying this is that, writing the rows of a 3×3 matrix in succession to give a vector

$$(a_{11} \ \ a_{12} \ \ a_{13} \ \ a_{21} \ \ a_{22} \ \ a_{23} \ \ a_{31} \ \ a_{32} \ \ a_{33}) \in \mathbf{R}^9,$$

the subspace of magic squares has dimension 3.

(iii) Verify that

$$E_1 = \begin{pmatrix} 0 & 1 & -1 \\ -1 & 0 & 1 \\ 1 & -1 & 0 \end{pmatrix}, \qquad E_2 = \begin{pmatrix} 1 & -1 & 0 \\ -1 & 0 & 1 \\ 0 & 1 & -1 \end{pmatrix},$$

$$E_3 = \begin{pmatrix} 1 & 1 & 1 \\ 1 & 1 & 1 \\ 1 & 1 & 1 \end{pmatrix}$$

are all magic squares. Why does it follow from (ii) that every 3×3 magic square A can be written uniquely in the form of a linear combination

$$A = \lambda_1 E_1 + \lambda_2 E_2 + \lambda_3 E_3$$

for $\lambda_1, \lambda_2, \lambda_3 \in \mathbf{R}$? (*Hint:* You are being asked to show that E_1, E_2, E_3 form a *basis* for the three-dimensional subspace of magic squares. Since there are three matrices E_i it is enough to show that they are *linearly independent*, considered as vectors in $\mathbf{R}^9$.) What are the λ_i when A is given by (8.1) above?

(iv) A 3×3 matrix has, in addition to its two 'main' diagonals, four other 'broken' diagonals:

$$a_{11} + a_{23} + a_{32}, \quad a_{12} + a_{21} + a_{33}, \quad a_{13} + a_{21} + a_{32}, \quad a_{12} + a_{23} + a_{31}.$$

If, for a magic square A, we require in addition all the broken diagonals to add up to the same magic constant r, then the square is called *pandiagonal*.

   In general, an $n \times n$ matrix has two main diagonals and $2n - 2$ broken diagonals, and if all these and all the rows and all the columns add up to the same r, then the matrix is a pandiagonal magic square.

   In the 3×3 case, expand your matrix M by the addition of four rows corresponding to the four broken diagonals and use the new matrix ($M1$ say) to show that the only pandiagonal magic squares are λE_3 for $\lambda \in \mathbf{R}$. (Thus you should find from the row reduced echelon form of $M1$ that only *one* entry of A is now arbitrary, and the others are all equal to this one.)

(v) We now look at *products* of 3×3 magic squares. By (iii) the product of two such squares has the form

$$(\lambda_1 E_1 + \lambda_2 E_2 + \lambda_3 E_3)(\mu_1 E_1 + \mu_2 E_2 + \mu_3 E_3). \tag{8.2}$$

Let

$$P = \begin{pmatrix} 0 & 0 & 1 \\ 0 & 1 & 0 \\ 1 & 0 & 0 \end{pmatrix}.$$

Verify that all products $E_i E_j$ (including $i = j$) can be written as linear combinations of I, P, E_3 (I being the 3×3 identity matrix), and hence that (8.2) is such a linear combination.

   Deduce that the product of an *even* number of 3×3 magic squares is a linear combination of I, P, E_3. Finally deduce that the product of an *odd* number of 3×3 magic squares is also magic. (For the 'even' case, using the result just proved, you need to show that multiplying together linear combinations of I, P, E_3 gives again a linear combination of these three matrices. This can be expressed by saying 'the subspace of $\mathbf{R}^9$

spanned by I, P, E_3 is closed under matrix multiplication'. Note this is *not* true of the subspace of magic squares itself. For the product of an *odd* number of magic squares you now know that this can be written as the product of a linear combination of I, P, E_3 and a linear combination of E_1, E_2, E_3.)

(vi) Explain why the condition that all the rows of a 3×3 matrix A add up to r is equivalent to the condition that $(1, 1, 1)$ is an eigenvector of A with corresponding eigenvalue r. Explain also why this last condition implies that, for any integer $k > 0$, $(1, 1, 1)$ is an eigenvector of A^k with eigenvalue r^k. (This is a standard property of eigenvectors and follows from the definition.) Now deduce from (v) that if A is a 3×3 magic square with magic constant r, and k is odd, then A^k is a magic square with magic constant r^k.

8.3 Magic squares size 4 × 4

(i) Set up a matrix corresponding to M above in the 4×4 case. It will be a 10×17 matrix. Use MATLAB to show that the rank is 9 and deduce that in $\mathbf{R}^{16}$ the 4×4 magic squares form a subspace of dimension 8.

(ii) Let $p = [p(1)\ p(2)\ p(3)\ p(4)]$ be a permutation of 1, 2, 3, 4, i.e. the numbers 1, 2, 3, 4 in some order. The *permutation matrix* A corresponding to p is the 4×4 matrix which is all zeros except for $A(i, p(i)) = 1$ for $i = 1, 2, 3, 4$. So, for example, if $p = [3\ 1\ 2\ 4]$ then

$$A = \begin{pmatrix} 0 & 0 & 1 & 0 \\ 1 & 0 & 0 & 0 \\ 0 & 1 & 0 & 0 \\ 0 & 0 & 0 & 1 \end{pmatrix}.$$

Note that this is in fact magic.

There are seven other permutations of 1, 2, 3, 4 which give magic squares in this way. Find them. As a hint, here is one fairly brute force way to enumerate the permutations of 1, 2, 3, 4 in MATLAB.

```
for a=1:4
  for b=1:4
    if b~=a
      for c=1:4
        if c~=a & c~=b
          for d=1:4
            if d~=a & d~=b & d~=c
p=[a b c d];

            end;
          end;
        end;
      end;
    end;
  end;
end;
```

After the permutation p is found, you will want to calculate the permutation matrix, then test whether it is magic. The quickest way to do this is to use the criterion $M\mathbf{v} = 0$ as in Question (i) in §8.2. Now, however, M will be the 10×17 matrix you have just found, and $\mathbf{v}$ will be a 17×1 column vector of the form $(-1 \ a_{11} \ \ldots \ a_{44})^{\top}$.

(iii) Add more rows to your matrix M to test for pandiagonal 4×4 magic squares (see Question (iv) in §8.2 for the definition). Call the resulting matrix $M1$. What is the dimension of the space now? Are any of these given by permutation matrices?

(iv) By finding the reduced row echelon form of your matrix $M1$, show that every 4×4 pandiagonal magic square has the form

$$\begin{pmatrix} a-b-c+d+e & -a+b+c+d+e & a+b-c-d+e & -a+b+3c+d-e \\ b+c-d+e & b+c+d-e & -b+c+d+e & b-c+d+e \\ -a+2c+2d & a-2c+2e & -a+2b+2c & a \\ 2b & 2c & 2d & 2e \end{pmatrix}$$

where a, b, c, d, e are arbitrary real numbers. Write an M-file to generate such matrices and use MATLAB to find the eigenvalues of several examples. Do you have any conjectures about the general result here? Can you explain *one* of the eigenvalues in the same way as Question (vi) in §8.2? (One relation between eigenvalues follows from the fact that the sum of the eigenvalues of any square matrix is equal to the *trace* of the matrix, that is, the sum of the entries in the leading diagonal—top left to bottom right.)

8.4 Magic squares size 5 × 5 (optional)

Find out what you can about the dimension of the space of 5 × 5 magic squares, and of the pandiagonal ones. Are the pandiagonal ones spanned in this case by permutation matrices? If so, find a basis consisting of permutation matrices. (You can find details in the article [16].)

9

GCDs, Pseudoprimes and Miller's Test

This chapter contains two investigations of a number-theoretic nature, building on the ideas of Chapter 3. Investigation A is on greatest common divisors (gcds), and B is on pseudoprimes and Miller's test for primality.

A GCDs of random pairs and triples of numbers

Aims of the project
The idea is to find, by theoretical and experimental means, estimates for the probability that a pair of randomly chosen positive integers have no common factor. This is also extended to triples of numbers.

Mathematical ideas used
Elementary ideas of probability are used (e.g., probability of two independent events is the product of their individual probabilities). There are several mathematical arguments given which need to be 'filled in' with some details—in fact, this project is more mathematical than computational. The idea is to use mathematical arguments and experiment to determine certain probabilities, such as: given three random integers, what is the probability that each of the three resulting *pairs* of integers are coprime—that is, have gcd 1 ?

MATLAB techniques used
The project uses the M-file `gcdiv.m` from Chapter 3, which calculates the gcd of two integers. It also requires the writing of some M-files which use loops and conditionals.

The M-file `gcdran.m` takes n pairs of 'random' numbers, each < 1000, and finds the gcd of each pair, then prints out the percentage of pairs with gcd equal to 1.

(i) Try running the M-file several times for say $n = 500$ and find the average percentage with gcd equal to 1.

(ii) Adapt the M-file gcdran to find the percentage which have gcd 2, and also the percentage with gcd 1 or 2 or 3. Thus for the latter case, for example $(4, 10), (6, 9), (12, 21)$ and $(6, 35)$ count towards those with gcd 1 or 2 or 3, whereas $(5, 10), (7, 35), (30, 50)$ do not.

(iii) By a bar-chart, histogram or other means, display (for a single run of say 1000 random pairs) the numbers of pairs which have gcds from 1 to 20. *Hint*: The best thing is to use a vector, say **v**, of length 20, whose k^{th} entry records the number of pairs with gcd equal to k. So you start by initialising

v=zeros(1,20)

and then, if a gcd is calculated to be $k < 21$ you increment

v(x) = v(x) + 1;

At the end, **bar(v)** will display the bar chart.

(iv) Consider the following 'argument', writing it out for yourself with any explanations which you feel able to offer.

Let x be the probability that a randomly chosen pair of numbers has gcd equal to 1, that is, the pair is coprime. Recall that for integers a, b, the symbol $a \mid b$ means 'a is a factor of b', that is, b/a is a whole number. Now

$$\gcd(a, b) = h \Leftrightarrow h \mid a \text{ and } h \mid b \text{ and } \gcd(a/h, b/h) = 1$$

so the probability that $\gcd(a, b) = h$ should be $(1/h)(1/h)x$. (Informally, the probability that h divides a random integer a is $1/h$.) Since every pair has *some* gcd we have

$$\sum_{h=1}^{\infty} \frac{x}{h^2} = 1.$$

Now $\sum_{h=1}^{\infty}(1/h^2) = \pi^2/6$. (This standard and remarkable result was proved by L. Euler around 1750. Proofs can be found in most books on complex analysis or Fourier series. Here, we simply take it as known.)

What does this give for x? How well does this accord with your experimental results? According to this, roughly what shape should your bar-chart/histogram for gcds up to 20 take?

(v) Extend the above argument (writing the new argument out for yourself) to suggest the probability y that three randomly chosen whole numbers a, b, c have $\gcd(a, b, c) = 1$. Adapt the M-file above (or your `gcdiv3.m` from Chapter 3) to find the number of random triples with gcd 1 and compare with the theoretical answer. If you need to know an approximate value for $\sum_{h=1}^{\infty}(1/h^3)$, then why not use MATLAB to calculate this sum to say 50 terms?

(vi) Here is an alternative 'argument' to that given in Question (iv) above. Write it out, filling in what details you can. Let p be a prime. The probability that a given integer a is divisible by p is $1/p$. So given *two* integers a, b the probability that they are *not both* divisible by p is $1-\frac{1}{p^2}$. Assuming these events, for different primes p, are all independent, the probability that there is *no* prime dividing both a and b is the product

$$\prod_{p}\left(1-\frac{1}{p^2}\right),$$

taken over all primes $p = 2, 3, 5, 7, 11, \ldots$. This is the probability that a and b are coprime.

An M-file called `primes.m` is available to you which produces a vector **p** containing the primes < 5000. (So $\mathbf{p}(1) = 2, \mathbf{p}(2) = 3$, etc.) Run this as usual by typing `primes` . How many such primes are there? Find the above product over the primes which are < 5000. (This will be a good approximation to the product over *all* primes.)

Does this agree with the probability x obtained in Question (iv)?

(vii) Take a random sample of triples of numbers a, b, c and find what percentage turn out to be *pairwise* coprime. Here, a, b, c are pairwise coprime if the gcds of a, b, of b, c and of a, c are all 1. Simply *cubing* the probability x that two numbers are coprime does not (you should find) give the probability that these three pairs are all coprime. Why do you think this is? *Warning*: To express '$x = y = z = 1$' in MATLAB do *not* write

`x==y==z==1`

since this statement will be 'true', that is, take the value 1, if, for example, $x = y$ and $z = 1$. Instead, write

`x==1 & y==1 & z==1.`

(viii) Here is a sketch of an 'argument' which suggests the probability of three numbers being pairwise coprime as above. Fill in what details you can. Let p be prime. The probability that *none* of the three numbers a, b, c is divisible by p is

$$\left(1-\frac{1}{p}\right)^3.$$

The probability that *exactly one* of a, b, c is divisible by p and the other two are not is

$$\frac{3}{p}\left(1-\frac{1}{p}\right)^2.$$

Hence the probability that *at most one* of the three is divisible by p is the sum of these two which after rearrangement becomes

$$\left(1-\frac{1}{p^2}\right)^2\left(1-\frac{1}{(p+1)^2}\right). \tag{9.1}$$

Now a, b and c are coprime if and only if, for *all* primes p, at most one of the numbers is divisible by p. Assuming these are all independent events the probability z that the three numbers are pairwise coprime is the product of the expressions (9.1) for *all* primes p. You should recognise the squared factor in front from Question (vi) above. Work out the value of the other factor using the same vector **p** of primes you used above. Now deduce the theoretical value of z.

Does this agree with the experimental value you found?

(ix) Consider the set of *all* pairs of numbers a, b where a and b are ≥ 2 and $\leq m$. For various m, what percentage of these are coprime? How close is this to the probability of random pairs being coprime?

B Pseudoprimes and Miller's test

Aims of the project
This project is based on the idea, introduced in Chapter 3, of *pseudo-prime*. Certain numbers 'masquerade as primes' and we shall explore one method commonly used to 'unmask' such numbers—to show that, in fact, they are not prime.

Mathematical ideas used
Fermat's theorem, given in §3.4, is used, and also the ability to work out remainders $\mathrm{rem}(a^n, m)$, where n and m may be fairly large numbers. This is done automatically with the MATLAB function **pow**. A relatively sophisticated primality test, Miller's test, is also introduced here. This can be used to 'unmask' some pseudoprimes, proving that they are composite without actually trying to factorise them.

MATLAB techniques used
The M-files for this project have been written for you, but you need

to amend one of them slightly to perform different tasks. The M-file gcdiv.m for calculating gcds is also used.

(i) An M-file called primes.m is available to you. This takes the numbers up to 5000 and decides for each one whether it is prime, putting the primes in a vector $\mathbf{p} = [\mathbf{p}(1), \ldots, \mathbf{p}(k)]$, where k is the number of primes involved. Thus $\mathbf{p}(1) = 2, \mathbf{p}(2) = 3, \mathbf{p}(3) = 5, \mathbf{p}(4) = 7, \mathbf{p}(5) = 11$, etc. So begin by typing primes. Once this has run, typing p will produce a long vector containing all these primes; say p(5) will produce the fifth prime, which is 11.

(ii) Recall from Chapter 3 that a *pseudoprime to base a* is a number n which is *not* prime, but which nevertheless satisfies $\text{rem}(a^{n-1}, n) = 1$. The significance of this is that every prime n automatically satisfies this relationship (so long as a is not a multiple of n), by Fermat's theorem (see Section 3.4). However, there is a very small minority of non-primes (composites) which 'masquerade as primes' to the extent that they, too, satisfy this relationship. One way of checking that a number n is not prime, without actually trying to factorise it—a hopelessly lengthy procedure for very large numbers—is to show that it fails to satisfy $\text{rem}(a^{n-1}, n) = 1$, for some a which is not a multiple of n. Pseudoprimes resist this attempt to show that they are not prime. Note that in this book we can only illustrate these ideas with relatively small numbers!

The M-file psp2.m finds pseudoprimes to base 2 which are $\leq m$ for a value of m input by the user. Note that we require the even number 2^{n-1} to leave remainder 1 when divided by n, say $2^{n-1} - kn = 1$. If n were even the left hand side of this equation would be even, which is a contradiction, so it is necessarily true that n is odd. To save some time in execution, the M-file checks only odd numbers n, starting at $n = 3$.

This M-file has comments to explain how it works. Take say $m = 2000$ and check that each number n output is indeed composite and that $2^{n-1} \equiv 1 \bmod n$, that is, 2^{n-1} leaves remainder 1 when it is divided by n. Of course you need to use

```
>> pow(2,n-1,n)
```

to do this.

Make amendments to psp2.m as necessary to find all numbers < 5000 which are:

(1) pseudoprimes to base 2,

(2) pseudoprimes to base 2 and 3,

(3) pseudoprimes to base 2, 3 and 5.

You should find that your list becomes steadily shorter: fewer and fewer numbers can masquerade as primes when several bases are used. For each pseudoprime, work out its factorisation into primes.

(iii) How many *primes* are there < 5000? (Use Question (i) above.) Suppose you are presented with an odd number n, where $1 < n < 5000$, and

$$\mathrm{rem}(a^{n-1}, n) = 1 \text{ for } a = 2, 3 \text{ and } 5.$$

What is the probability that n is not a prime? The moral here is that n is 'very likely' to be prime.

(iv) If you want to find the numbers n which are pseudoprimes just to base 3, rather than simultaneously to bases 2 and 3, then you should also consider even numbers n. The same holds for base 5. Are there in fact any even pseudoprimes n to base 3 or base 5 which are less than 10000?

(v) **Miller's test** If a number n has $\mathrm{rem}(a^{n-1}, n) \neq 1$, (that is, $a^{n-1} \not\equiv 1 \bmod n$) for some a with n not dividing a, then we know from Fermat's theorem that n is *not* prime. There is a more sophisticated method for flushing out composite (that is, not prime†) numbers, called *Miller's test*, which was published in 1976. For more details of the theory of Miller's test see, for example, [7], Ch.5.

Here is the test. Start with an odd $n > 1$, and a number b (the *base*), which is coprime to n, that is, $\gcd(b, n) = 1$. Then carry through the following steps:

Step 1 Let $k = n - 1, r = \mathrm{rem}(b^k, n)$. If $r \neq 1$ then Miller's test to base b has been *failed*, otherwise continue. (Note that if n passes Step 1, then n is either prime or a pseudoprime to base b.)

While k is even and $r = 1$, repeat the following

Step 2 Replace k by $k/2$, then replace r by the new value of $\mathrm{rem}(b^k, n)$.

When k becomes odd or $r \neq 1$:

if $r = 1$ or $r = n - 1$ then n has *passed* Miller's test to base b;

if $r \neq 1$ and $r \neq n - 1$ then n has *failed* Miller's test to base b.

† For technical reasons, the number 1 is regarded as neither prime nor composite. This need not worry us in the slightest!

The important result is that *if n is prime then n always passes the test*. Thus if n fails the test then it is composite. Note that failing Step 1 amounts to failing to satisfy Fermat's theorem.

The M-file `miller.m` implements this test. See if you can understand how it does this. Use the M-file to do the following:

(vi) Test `miller.m` by using some primes, which must pass the test. Write out the residues obtained. You can get plenty of primes to try from the result of Question (i) above.

(vii) Check that $n = 1373653$ passes the test for $b = 2$ and $b = 3$ but fails it for $b = 5$; write down the results of doing the test for these b. Is this n prime or composite? Let $c = \text{rem}(5^{n-1}, n)$. (Note that this *cannot* be evaluated by the MATLAB function `rem` since the numbers are far too big. But the M-file `miller.m` *tells you* what c is !) Find $\gcd(c - 1, n)$ and check that this is a *proper* factor of n, that is, a factor which isn't 1 or n.

(viii) Check that $n = 25326001$ passes Miller's test to bases 2, 3 and 5 but fails it for base 7 (write out the residues occurring in the tests). Let $c = \text{rem}(7^{n-1}, n)$. Find $\gcd(c - 1, n)$ and check that it is a proper factor of n.

(ix) Find the first number $> 5 \times 10^7$ which could possibly be prime. Write down exactly what you do to find this number. (So the fewer the times you need to use Miller's test the better. For example, do not test 50000002 for primality by Miller's test!)

(x) Which of the pseudoprimes found above (Question (ii)) are unmasked as composites by using Miller's test to base 2?

(xi) There is a composite number between 2000 and 3000 which passes Miller's test to base 2. Find it. Does it pass Miller's test to base 3?

(xii) A tiny bit of theory. The notation $x \equiv y \bmod n$ ('x is congruent to $y \bmod n$') means that $n \mid (x - y)$ (that is, n divides exactly into $x - y$). Thus, if $r = \text{rem}(a, n)$, then $a \equiv r \bmod n$. You may assume the standard properties (which are easy to prove): If $x \equiv y$ and $u \equiv v \bmod n$ then $x \pm u \equiv y \pm v$ and $xu \equiv yv \bmod n$. In particular, $x^2 \equiv y^2 \bmod n$.

Now suppose that n is odd and $2^{r+1} \mid (n - 1)$, where r is an integer > 0, and that $\gcd(b, n) = 1$. Suppose that

$$b^{(n-1)/2^r} \equiv 1 \bmod n, \qquad\qquad (9.2)$$

$$c = b^{(n-1)/2^{r+1}} \not\equiv \pm 1 \bmod n. \qquad\qquad (9.3)$$

Use (9.2) and successive squaring to show that n is a pseudoprime to base b. Show also that n fails Miller's test to base b, and that $n \mid (c^2 - 1)$, using (9.3).

Let $h = \gcd(c - 1, n)$. The claim is that $h \neq 1$ and $h \neq n$, that is, that h is a *proper* factor of n. Show this as follows. (a) If $h = 1$, deduce $n \mid (c + 1)$. (You may use the standard result $x \mid yz, \gcd(x, y) = 1, \Rightarrow x \mid z$.) Why is this a contradiction? (b) If $k = n$ then deduce $n \mid (c - 1)$. Why is this a contradiction?

Check that 10004681 is a pseudoprime to base 2 which fails Miller's test to base 2. Factorise it using the above method. (Ideally you should end up with the prime factors of 10004681. Remember that the vector **p** of primes is available to you.)

For the pseudoprimes to base 2 which you found in (a) above, check which ones can be factorised by the above method. Write down the steps you go through in testing the numbers, and list those for which the method does not work.

10
Graphics: Curves and Envelopes

This chapter contains three graphics investigations. Investigation A is about certain special cases of curves obtained by rolling one circle on another circle; B is a further exploration of envelopes of families of lines, introduced in Chapter 4, and C is on curves which have 'constant width' in the sense that all pairs of parallel tangents are the same distance apart.

A Rose curves and epicycloids

Aims of the project
Rose curves are curves with 'petals'—they don't particularly resemble roses!—and we shall explore these theoretically and experimentally. (Some more ideas for project work can be found in [8].) Then we shall take a certain envelope defined by chords of a circle and relate it to a curve obtained by rolling one circle on another.

Mathematical ideas used
The project is about parametrised plane curves, including polar coordinates, and envelopes of lines (§4.6). Trigonometrical formulae come into some of the calculations.

MATLAB techniques used
You will need to use some M-files from Chapter 4 to draw curves and envelopes of lines.

Rose curves These are the special case of hypocycloids (see Exercise 4.5), where $a = 2m/(m+1), b = 1$ and $d = a - 1$. Here, m is an integer > 0. Figure 10.1 shows the example $m = 4$.

Make an investigation of these curves. Use the M-file `hypocy.m` . In particular investigate experimentally the following numbers, expressed in terms of m: (i) the smallest value of t which makes the curve close

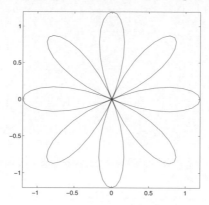

Fig. 10.1. A rose curve with $m = 4$ and eight 'petals'.

up and start to repeat itself; (ii) the number of times the curve passes through the origin before it closes up and starts to repeat itself; (iii) the number of 'petals' the rose has.

Then do a mathematical investigation as follows. Put $t = (m-1)\theta$ in the parametrisation of the hypocycloid given in equation (4.2). Then use the famous formulae

$$\cos\alpha + \cos\beta = 2\cos\left(\frac{\alpha+\beta}{2}\right)\cos\left(\frac{\alpha-\beta}{2}\right),$$

$$\sin\alpha - \sin\beta = 2\cos\left(\frac{\alpha+\beta}{2}\right)\sin\left(\frac{\alpha-\beta}{2}\right),$$

to show that the rose curve can be written as $r = 2\frac{m-1}{m+1}\cos(m\theta)$ in polar coordinates r, θ. (This just means that $x = r\cos\theta, y = r\sin\theta$.) You can assume that a polar curve $r = f(\theta)$, where f is some function, will close up and start to repeat (a) when $\theta = \pi$ if $f(\theta+\pi) = -f(\theta)$ for all θ, and (b) when $\theta = 2\pi$ if $f(\theta+\pi) \neq -f(\theta)$ but $f(\theta+2\pi) = f(\theta)$ for all θ. Use this to find the answers to (i), (ii), (iii) above, and compare with your experimental answers.

Epicycloids

(i) Show that the equation of the chord of the circle $x^2 + y^2 = 1$ joining $(\cos t, \sin t)$ to $(\cos(mt), \sin(mt))$ is

$$x(\sin t - \sin mt) - y(\cos t - \cos mt) + \sin(m-1)t = 0. \qquad (10.1)$$

(You will need to remember the trigonometrical formula for $\sin(a-b)$.)

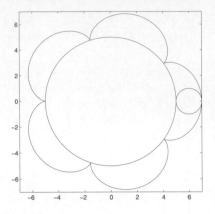

Fig. 10.2. An epicycloid obtained by rolling a circle radius 1 outside a circle radius 5.

Use `linenv.m` to draw the envelope of these chords for some values of m, say 2, 3 and 4. Print *one* of these out.

(ii) Verify that the point

$$x(t) = \frac{m\cos t + \cos(mt)}{m+1}, y(t) = \frac{m\sin t + \sin(mt)}{m+1}$$

satisfies both equation (10.1) and the equation obtained by differentiating equation (10.1) with respect to t. This is a special case of the situation of Exercise 4.7, and shows that the above point $(x(t), y(t))$ gives a parametrisation of the envelope of the chords. That is, the curve which your eye picks out from the drawing of all the chords, where the lines appear to cluster, is parametrised as above.

(iii) An *epicycloid* is similar to a hypocycloid (Exercise 4.5) but the rolling circle rolls *outside* the fixed one. The parametrisation is (using u for a reason which will become clear)

$$x = (a+b)\cos\left(\frac{bu}{a+b}\right) + d\cos u, \quad y = (a+b)\sin\left(\frac{bu}{a+b}\right) + d\sin u.$$

Figure 10.2 shows the example $a = 5, b = 1, d = 1$, with the fixed circle also drawn and the rolling circle in its position for $u = 0$.

Show (mathematically!) that by putting $u = mt$ and taking suitable values of a, b, d in terms of m, this epicycloid can be made to coincide with the above parametrization of the envelope of chords. You need, for example, $m/(m+1) = a+b$, $1/(m+1) = d$.

(iv) Amend `hypocy.m` so that it draws epicycloids. *Note*: You will need to make one subtle amendment, namely 'upper' should be redefined as `abs(a+b)+abs(d)`. If you name the parameter as u then you also need to use `u=tl:tstep:tu` instead of `t=tl:tstep:tu`. There is no need to rename `tl` and `tu`. The rest of the changes are simply in the x and y lines. Draw some of these epicycloids for small (integral) values of m and compare with the envelope of chords.

(v) Prove the formula for the parametrisation of an epicycloid (in Question (iii) above).

B Envelopes

Aims of the project
The idea is to look at two situations where a family of lines creates an envelope. One is a family of perpendicular bisectors similar to that encountered in Chapter 4 (Exercise 4.6), and the other comes from a sliding ladder. The investigation is both mathematical and experimental.

Mathematical ideas used
These include parametric equations of curves, elimination of a variable between two equations, tangents to curves and simple trigonometry. The ideas on envelopes come from Chapter 4.

MATLAB techniques used
The first part of the investigation uses the M-file `linenv.m` to draw envelopes of perpendicular bisectors. You also need to draw an additional curve on the same figure. The second part involves the solution of polynomial equations, drawing parametric curves and drawing envelopes of lines.

Perpendicular bisectors
Investigate the envelope of perpendicular bisectors of lines joining the point $(a, 0)$ (where you can assume $a \geq 0$) to the points of the unit circle $x^2 + y^2 = 1$, which is parametrised by $(\cos t, \sin t)$. Thus to begin with, show that the equation of the perpendicular bisector is

$$2x(a - \cos t) - 2y \sin t + 1 - a^2 = 0 \qquad (10.2)$$

(see Exercise 4.6). Then use `linenv.m` to draw the envelope for various values of a. The M-file should have a couple of lines added to it so that it draws this circle as well, in red. Print out one example for $a > 1$ and one for $a < 1$.

The rest of this part is a mathematical investigation to explain the pictures just obtained. For the perpendicular bisectors given by equation (10.2) use the method of Exercise 4.7, to show that the equation of the envelope of perpendicular bisectors is, for $a \neq 1$,

$$4x^2 - 4ax + \frac{4y^2}{1 - a^2} = 1 - a^2.$$

So you have to *eliminate* t between two equations, namely equation (10.2) and the derivative of this equation with respect to t. One way to eliminate t is to arrange the two equations as equations for $\sin t$ and $\cos t$, solve for these two, and then use $\sin^2 t + \cos^2 t = 1$.

What is the envelope if $a = 1$? Is there a simple geometrical explanation for this?

Assume $a \neq 1$. Make the substitution $X = x - \frac{a}{2}$ to reduce the equation of the envelope to the form

$$4X^2 + \frac{4y^2}{1 - a^2} = 1.$$

What kind of curve is this? Explain with examples how it fits with the pictures.

Can you find any significance in the particular perpendicular bisectors which arise when the line joining $(a, 0)$ to the circle is *tangent* to the circle? Of course this requires $a > 1$. (*Hint*: Show that these correspond with $\cos t = 1/a$ and have equations $X\sqrt{a^2 - 1} = \pm y$.)

Sliding ladders

A ladder of length l rests on the ground and on a vertical wall, just also resting on a rectangular box of sides a and b, as in Figure 10.3, left.

Writing x for the length shown, show that $f(x) = 0$, where

$$f(x) = x^4 - 2ax^3 + (a^2 + b^2 - l^2)x^2 + 2al^2x - a^2l^2.$$

Take $l = 10$ and $a = 2$. Verify that $b = 5$ gives a physically possible solution (that is, the equation has a root x which is physically possible), but $b = 6$ does not. Illustrate with the graph of f but also use `roots` to find the roots numerically. The M-file `paramc.m` draws parametrised curves $(x, y) = (x(t), y(t))$. In the special case of a function graph $y = f(x)$ you can take $x = t, y = f(t)$, so you have to insert these in the correct place in `paramc.m`. (Do not confuse the y here with the y in either of the diagrams! The M-file is a general purpose parametric curve plotter, so x, y are the most natural variables to use in it.)

Take $a = 2$ and $b = 5$ and find all solutions when $l = 20$. Try also

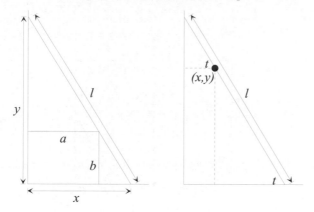

Fig. 10.3. Left: a ladder of length l leaning against a vertical wall and resting on a box of dimensions a and b. Right: finding the equation of the ladder. Note that x, y have different meanings in the two diagrams!

$a = 2, b = 5, l = 9.582299$. What do you notice here? What happens if l is decreased or increased slightly?

We now take the box away and imagine the ladder sliding down the wall, keeping in a vertical plane (a frightening possibility!) and find the envelope of all the ladder positions.

Show that the equation of the straight line along the ladder, given the angle t in the right-hand diagram of Figure 10.3, is

$$\frac{x}{\cos t} + \frac{y}{\sin t} = l.$$

Show from the method of Exercise 4.7 that the envelope of the ladder lines is given by $x = l\cos^3 t, y = l\sin^3 t$. Now eliminate t to show that the equation of the envelope is

$$x^{2/3} + y^{2/3} = l^{2/3}.$$

Also use `linenv.m` to draw the envelope of the ladder lines.

Show that the mid-point of the ladder describes part of a circle as the ladder slides. Can you include this on the picture of the envelope? Can you see a connection with up-and-over garage doors?

Why can you deduce from the equation of the envelope that the condition for a physically possible solution of the original ladder and box problem to exist is

$$a^{2/3} + b^{2/3} \le l^{2/3}?$$

Check this with the examples above.

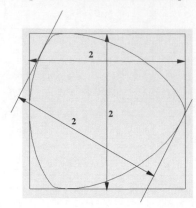

Fig. 10.4. All widths, measured between parallel tangents, are equal to 2.

C Curves of constant width

Aims of the project
The circle is by no means the only curve which has 'constant width' in the
sense that all pairs of parallel tangents are the same distance apart. This
project investigates properties of curves of constant width, including
how to make them, and how they are related to other geometrical ideas
such as envelopes of lines. (There are some interesting observations on
curves—and surfaces—of constant width in [5].)

Mathematical ideas used
Equations of lines and tangents, linear equations and 3×3 determinants;
envelopes of lines (Chapter 4) come in too.

MATLAB techniques used
Drawing parametric curves and envelopes of lines, using given M-files.

This project is about curves of constant width. The *width* of a curve is
measured between a pair of parallel tangents and *constant width* means
that all such widths are equal. Surprisingly, the circle is not the only
curve with this property. See Figure 10.4. Curves of constant width
have been applied in very practical situations, for example the Wänkel
car engine and bits for drilling (nearly) square holes. Coins have to be
curves of constant width to work in coin-operated machines; the British
20p and 50p coins are noncircular curves of constant width.

We shall generally only consider curves that are 'convex'; this implies

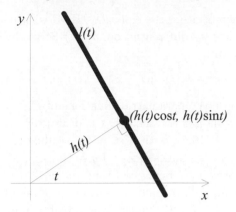

Fig. 10.5. 'Pedal coordinates' of a line.

that there are exactly two tangents parallel to any given direction. We shall meet some more 'singular' examples later.

See §4.6 and Exercise 4.7 for details of envelopes.

The 'pedal' construction

Let h be a function of the angle t. In Figure 10.5, $l(t)$ is the line through the point $(h(t) \cos t, h(t) \sin t)$, perpendicular to the direction from this point to the origin.

(i) Show that the equation of $l(t)$ is

$$x \cos t + y \sin t = h(t). \tag{10.3}$$

(ii) Let $h(t) = 1 + \frac{1}{8} \cos(3t)$. Amend the M-file `linenv.m` so that it draws the envelope E of the lines $l(t)$ for $0 \le t \le 2\pi$. Print this envelope out, taking care to adjust the limits of x and y so that the whole envelope E is clearly visible on the picture.

(iii) The lines $l(t)$ are of course all *tangent* to the envelope E drawn in (ii). Check that $h(t) + h(t + \pi) = 2$ for all t with the function h given there. Why does this imply that the pairs of parallel tangents to E are all a distance 2 apart, and hence that E is a curve of constant width 2?

(iv) Recall from Exercise 4.7 that to find equations for the envelope we take the equation (10.3), together with the equation

$$-x \sin t + y \cos t = h'(t) \tag{10.4}$$

obtained by differentiating the equation (10.3) with respect to t. (Here and below, $'$ stands for differentiation, d/dt.) Solve the equations (10.3) and (10.4) to obtain

$$x(t) = h(t)\cos t - h'(t)\sin t \quad y(t) = h(t)\sin t + h'(t)\cos t, \qquad (10.5)$$

which are therefore the parametric equations of E.

Amend the M-file `paramc.m` so that it draws the parametrised curve given by equations (10.5), with h as in (ii), calling the amended M-file `constw1.m`. Adjust the limits of x and y so that E *just* fits in the square on the screen (of course, being constant width, it *will* just fit inside a square!), and print out this picture. The curve E should match up with the one you printed out in (ii), apart from possibly different scaling. Join (by hand) the points where your curve touches the top and bottom of the square, and the points where it touches the two sides of the square. Do you notice anything special about these two lines?

(v) Explain why, changing $h(t)$ (as in (ii)) by adding a constant a, another curve of constant width is obtained. What is the width of this curve? Amend your M-file `constw1.m` so that it draws this curve for each of the eleven values $a = 0, 0.1, 0.2, \ldots, 1.0$ on the same picture, calling the resulting M-file `constw2.m`. Remember that using `hold on` between the drawing of the various curves keeps them all on the same picture. Use `hold off` at the end of the M-file. Adjust the limits of x and y so that the largest of the eleven curves fits exactly on the square of the screen and print out this picture.

(vi) What happens if you take $a = -0.5$ in (v)? Draw the resulting 'curve' on the screen and make a hand sketch of it. Do you think it has constant width?

(vii) As already noted, the function h in (ii) satisfies

$$h(t) + h(t + \pi) = k, \qquad (10.6)$$

where k is constant (actually $k = 2$ in this example). Another way of satisfying the equation (10.6) is to make h itself constant. What curve E is produced by the equation (10.5) in that case? (You can answer that without using the computer!) Choose a *nonconstant* function h which satisfies the equation (10.6) and produces a curve E different from that studied above. Print out the curve E as given by your h and the equation (10.5), choosing scales so that it fits exactly on the square of the picture.

(viii) Let h be any function satisfying the equation (10.6), and let E be

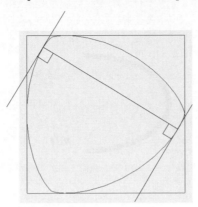

Fig. 10.6. The line joining the two points of contact of parallel tangents is in fact normal to the curve at both contact points.

the envelope parametrised as before by equations (10.5). Show that

$$x(t + \pi) = x(t) - k \cos t, \quad y(t + \pi) = y(t) - k \sin t.$$

Deduce from this that the line joining the points of the envelope E at t and $t + \pi$ is always perpendicular to the lines $l(t)$ and $l(t + \pi)$ (see Figure 10.6).

Completing a given 'half-curve'

Here, we start with a curve which has parallel tangents at its endpoints but nowhere else. We shall take the example of the half-ellipse in Figure 10.7. Here, the two parallel tangents are a distance 2 apart and we construct a curve of constant width 2. The half-ellipse is parametrised by

$$x = a \cos t, \quad y = \sin t, \quad \text{for} \quad \frac{\pi}{2} \leq t \leq \frac{3\pi}{2}.$$

For each tangent to the half-ellipse, say at $(a \cos t, \sin t)$, we take the line $m(t)$ parallel to this tangent, at distance 2 (see Figure 10.7).

(ix) Show that the line $m(t)$ has equation

$$x \cos t + ya \sin t = a - 2\sqrt{a^2 \sin^2 t + \cos^2 t}. \tag{10.7}$$

(x) The equation (10.7) is not in quite the same form as equation (10.3), because of the presence of a in the coefficient of y. However, the method of solving for the points of the envelope E of the lines $m(t)$ is still the

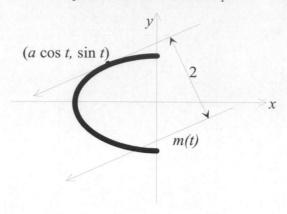

Fig. 10.7. A curve of constant width based on half an ellipse.

same as before. Writing $r(t)$ for the right-hand side of equation (10.7), show that the points of the envelope are

$$x(t) = r(t)\cos t - r'(t)\sin t, \quad y(t) = \frac{1}{a}(r(t)\sin t + r'(t)\cos t). \quad (10.8)$$

Why does the envelope E, together with the half-ellipse, make a curve of constant width 2?

(xi) Amend your M-file `constw1.m`, or write another M-file, to draw the envelope curve E given by equation (10.8) for $\pi/2 \le t \le 3\pi/2$, and also to draw the original ellipse for the same range of values of t. Call the result `constw3.m`.

(xii) Draw the following three examples of E:

$$a = 0.75, \quad xl = -1, \quad xu = 2, \quad yl = -1.5;$$
$$a = 1.25, \quad xl = -1.25, \quad xu = 1.25, \quad yl = -1.25;$$
$$a = 3, \quad xl = -3, \quad xu = 1, \quad yl = -2.$$

As usual xl, xu, yl stand for the lower and upper limits of x and the lower limit of y. In each case make a *hand sketch* of the half-ellipse and envelope. You should find that the third curve is *singular*, that is, has *cusps* or sharp points.

(xiii) By taking various values of a (and suitable limits for x and y) determine as accurately as you can the range of values of a for which the envelope E above is free from cusps. You need only consider $a > 0$.

(xiv) The remaining work here uses the theory of envelopes to find the range of values of $a > 0$ for which there are *no* cusps on the above envelope. Writing equation (10.7) as

$$px + qy = r,$$

it can be shown that the condition for a singular point (cusp) to exist is that (using $'$ for differentiation as before)

$$\begin{vmatrix} p & q & r \\ p' & q' & r' \\ p'' & q'' & r'' \end{vmatrix} = 0$$

has a solution t between $\pi/2$ and $3\pi/2$. Using $p = \cos t, q = a\sin t$ show that adding the top row to the bottom row and evaluating the determinant by the bottom row reduces this condition to simply $r + r'' = 0$ (remember that a is nonzero). Hence show that the condition is $v = 0$, where

$$v = -2u + a + \frac{2(a^2 - 1)^2 \sin^2 t \cos^2 t}{u^3} - \frac{2(a^2 - 1)(\cos^2 t - \sin^2 t)}{u}$$

$$\text{and} \quad u = \sqrt{a^2 \sin^2 t + \cos^2 t}.$$

Write a (short) M-file to draw the graph of v:
```
t=-pi/2:0.01:3*pi/2;
v= above expression;
plot(t,v)
hold on
plot([pi/2,3*pi/2],[0,0])
hold off
```
What does the second `plot` statement do?

Draw a *hand sketch* of the resulting graph of v for the three values $a = 0.75, 1.25, 3$ and verify that in these cases the curve in (xi) above is free from cusps precisely when the graph of v fails to cross the axis.

What is the range of values of a, according to this method, for which the envelope E is free from cusps?

11

Zigzags and Fast Curves

This chapter contains two investigations which use graphics: A, on spirographs and zigzags, and B, on the problem of determining the shape of a wire which gives the fastest time of descent for a bead sliding down the wire.

A Spirographs and zigzags

Aims of the project

The idea here is to draw a zigzag line determined by a simple rule. The interest lies in determining how many times the line must zig and zag before it closes up, and in working out how large the resulting picture is, so that it can be drawn fitting neatly on the screen. We shall also investigate the connection between the zigzags and certain spirograph curves (epicycloids). The basic idea for this project comes from the book [1].

Mathematical ideas used

Vectors in the plane, linear equations and 2×2 matrices, regular polygons, gcds, trigonometric formulae and ellipses all come into this project.

MATLAB techniques used

This is a project about drawing sequences of lines. You need to amend a given program to make it do successively more tasks automatically. One of these involves calculating gcds, and another involves setting the screen size to fit the zigzag.

Construction of the zigzag

This project involves drawing zigzag lines according to a simple rule, and examining the mathematics behind these constructions. There is an intimate connection between the zigzags and certain *epicycloids*, which are

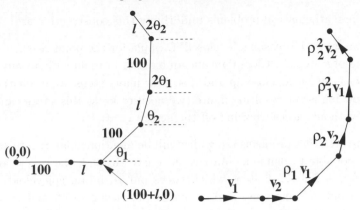

Fig. 11.1. The basic zigzag, defined by lengths of 100 and l, and angles θ_1, θ_2. In the right-hand figure, ρ_1 and ρ_2 are rotations through θ_1 and θ_2 respectively.

obtained by rolling one circle on another and drawing the path traced out by a point rigidly attached to the rolling circle. (Compare Investigation A of Chapter 10.)

The basic idea is illustrated in Figure 11.1. A straight horizontal line is drawn to the right from the origin, of length 100. At the end of this another straight horizontal line is drawn of length l. If $l < 0$ then the line is drawn to the left and otherwise to the right; in either case it terminates at $(100 + l, 0)$. At this stage we say that one *step* has been completed.

Now the zigzagging begins. We have two angles θ_1, θ_2 given to us (usually they will be whole numbers of *degrees*). We draw a straight line of length 100 from $(100+l, 0)$, at an angle θ_1 with the positive x-axis (so this angle is measured anticlockwise from this axis). The termination of this line is at $(100 + l + 100\cos\theta_1, 100\sin\theta_1)$. From this point we draw a line of length l at an angle θ_2 with the horizontal, thereby arriving at the point

$$(100 + l + 100\cos\theta_1 + l\cos\theta_2, 100\sin\theta_1 + l\sin\theta_2).$$

At this stage, two *steps* have been completed.

The lengths of the added lines are always alternately 100 and l, but the *angles* between the added lines and the horizontal go up by θ_1 and θ_2 respectively at every step. Thus the third step consists of two lines at angles of $2\theta_1, 2\theta_2$ to the horizontal, the fourth step of two lines at angles of $3\theta_1, 3\theta_2$ to the horizontal, etc.

The mathematical problems underlying this construction are:

- Can we find a reasonable 'closed' formula for the point reached at the end of k steps? ('Closed' means an explicit formula without any ...s.)
- Does the zigzag close up and if so how many steps are needed?
- How big is the resulting figure (we need to know this so we can draw it without pieces whizzing off the screen).

There are other problems too which will be mentioned later: for example it is possible to define a 'spirograph' curve through the points arrived at after $1, 2, 3, \ldots, k$ steps which turns out to be an epicycloid. This epicycloid in some sense approximates the zigzag itself, and there are examples where the epicycloid is a simple curve like an ellipse.

The M-file **zz1.m** draws the zigzag but it requires you to give a lot of help, by specifying the number of steps to be executed and the size of the graphics window ($xl \leq x \leq xu, yl \leq y \leq yu$). In due course you will have amended this so that the number of steps to *closure* is calculated in advance, and the size of the window set automatically. The one thing **zz1.m** does do for you is to make the graphics window square, that is, $xu - xl = yu - yl$. In fact yu is put equal to $yl + xu - xl$.

As a start, try running **zz1.m** with the inputs $l = 40$, $\theta_1 = 45$, $\theta_2 = 9$, steps $= 40$, $xl = -320$, $xu = 460$, $yl = -15$. Try also $l = 50$, $\theta_1 = 175$, $\theta_2 = 185$, steps $= 72$, $xl = 0$, $xu = 150$, $yl = -74$.

A vector formula for the point reached after k steps

For consistency of notation we shall replace 100 by l_1 and l by l_2 in what follows. Let $\mathbf{v}_1, \mathbf{v}_2$ be horizontal vectors of lengths l_1, l_2 respectively, and let ρ_1, ρ_2 denote rotations anticlockwise through θ_1, θ_2 respectively. Thus for example if $\mathbf{v}$ is the vector (a, b) then the effect of ρ_1 on $\mathbf{v}$ is to turn it into the vector

$$\begin{pmatrix} \cos\theta_1 & -\sin\theta_1 \\ \sin\theta_1 & \cos\theta_1 \end{pmatrix} \begin{pmatrix} a \\ b \end{pmatrix} = \begin{pmatrix} a\cos\theta_1 - b\sin\theta_1 \\ a\sin\theta_1 + b\cos\theta_1 \end{pmatrix}. \tag{11.1}$$

We can write ρ_1^k for the rotation through $k\theta_1$; the matrix for this is obtained by replacing θ_1 by $k\theta_1$. For ρ_2, we replace θ_1 by θ_2. Using this notation we can write the position of the point at the end of k steps of the zigzag as (see the right-hand part of Figure 11.1):

$$\mathbf{v}_1 + \mathbf{v}_2 + \rho_1\mathbf{v}_1 + \rho_2\mathbf{v}_2 + \rho_1^2\mathbf{v}_1 + \rho_2^2\mathbf{v}_2 + \ldots \rho_1^{k-1}\mathbf{v}_1 + \rho_2^{k-1}\mathbf{v}_2$$
$$= (\mathbf{v}_1 + \rho_1\mathbf{v}_1 + \rho_1^2\mathbf{v}_1 + \ldots \rho_1^{k-1}\mathbf{v}_1)$$
$$+ (\mathbf{v}_2 + \rho_2\mathbf{v}_2 + \rho_2^2\mathbf{v}_2 + \ldots \rho_2^{k-1}\mathbf{v}_2). \tag{11.2}$$

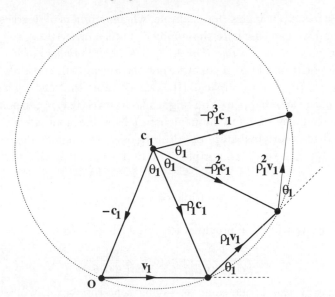

Fig. 11.2. Part of a 'regular polygon' with central angle θ_1 and centre c_1.

The two brackets in equation (11.2) can be rewritten in a much more convenient form. In fact the vectors occurring in the first bracket are simply the vectors along the sides of a *regular polygon* with side equal to the length l_1 of $\mathbf{v}_1$ as shown in Figure 11.2. The point $\mathbf{c}_1$ is the centre of the polygon and the distances from $\mathbf{c}_1$ to the various vertices (starting at $\mathbf{O}$) are all the same. *Explain the labels on the angles and sides in the figure and the following two equations.*

$$\mathbf{v}_1 = \mathbf{c}_1 - \rho_1\mathbf{c}_1, \qquad (11.3)$$

$$\mathbf{v}_1 + \rho_1\mathbf{v}_1 + \rho_1^2\mathbf{v}_1 + \ldots + \rho_1^{k-1}\mathbf{v}_1 = \mathbf{c}_1 - \rho_1^k\mathbf{c}_1. \qquad (11.4)$$

Hence the point of the zigzag reached after k steps is, using equations (11.2), (11.4) and the corresponding result about $\mathbf{v}_2$,

$$\mathbf{c}_1 - \rho_1^k\mathbf{c}_1 + \mathbf{c}_2 - \rho_2^k\mathbf{c}_2. \qquad (11.5)$$

So to find a good expression for the point that the zigzag has reached we just need to find $\mathbf{c}_1$ and $\mathbf{c}_2$. This is easy from equation (11.3) and its counterpart for $\mathbf{v}_2$, but first we look at the number of steps.

The number of steps to closure
It is not obvious that the zigzag *ever* closes, of course. But so long as

we stick to simple angles (for example, whole numbers of degrees) it *will* close, and we can calculate the number of steps needed.

From now on we assume that θ_1, θ_2 are whole numbers of degrees.

From equation (11.5) it follows that the zigzag will certainly close if both $\mathbf{c}_1 = \rho_1^k \mathbf{c}_1$ *and* $\mathbf{c}_2 = \rho_2^k \mathbf{c}_2$. (It is conceivable that the zigzag closes before both these happen. You might like to investigate that possibility! But we shall say no more about it here.) Now if a nonzero vector $\mathbf{c}_1$ is returned to its original state by k rotations ρ_1 through θ_1 degrees, then it must be that $k\theta_1$ is a multiple of $360°$. Explain why this is the same as

$$k = \text{ multiple of } \frac{360}{\gcd(360, \theta_1)}.$$

So the zigzag will close provided k is a multiple of *both*

$$\frac{360}{\gcd(360, \theta_1)} \text{ and } \frac{360}{\gcd(360, \theta_2)}.$$

Explain why this is the same as saying that the zigzag will definitely close when $k = s$, where

$$s = \frac{360}{\gcd(360, \theta_1, \theta_2)}. \tag{11.6}$$

(Thus you need to show s is a common multiple of the two numbers given above. In fact it is the *least* common multiple. .)

Use equation (11.6) to calculate the quantity 'steps'—that is, the number of steps to closure—in `zz1.m`, using the available M-file `gcdiv.m` for calculating gcds. Call the resulting M-file `zz2.m`. Check several examples to see that the zigzag does close. Write out the data which you input to the M-file, and the calculated number of steps to closure.

Size of the zigzag

We want to calculate in advance the size of the zigzag so that the graphics window can be set accordingly. In terms of equation (11.5) we shall take $\mathbf{c} = \mathbf{c}_1 + \mathbf{c}_2$ at the centre of the screen, i.e. at $(\frac{1}{2}(xl + xu), \frac{1}{2}(yl + yu))$. So we presumably need to find $\mathbf{c}$ and also to find the other terms of the equation (11.5).

Finding the 'centre' c

Show from equation (11.3), the matrix for the rotation ρ_1 given in equation (11.1), and the fact that $\mathbf{v}_1 = (l_1, 0)^\top$ (as a column vector) that

$$\mathbf{c}_1 = \begin{pmatrix} 1 - \cos\theta_1 & \sin\theta_1 \\ -\sin\theta_1 & 1 - \cos\theta_1 \end{pmatrix}^{-1} \mathbf{v}_1$$

$$= \frac{l_1}{2\sin\frac{1}{2}\theta_1}\left(\sin\frac{1}{2}\theta_1, \cos\frac{1}{2}\theta_1\right)^{\mathsf{T}}.$$

Hint: You will need to use the 'half-angle formulae'

$$1 - \cos\theta_1 = 2\sin^2\frac{1}{2}\theta_1, \sin\theta_1 = 2\sin\frac{1}{2}\theta_1\cos\frac{1}{2}\theta_1.$$

Of course a similar formula holds for c_2, all suffix 1s being replaced by suffix 2s.

Now use the matrix for the rotation ρ_1^k through $k\theta_1$, namely

$$\begin{pmatrix} \cos k\theta_1 & -\sin k\theta_1 \\ \sin k\theta_1 & \cos k\theta_1 \end{pmatrix},$$

a similar matrix for ρ_2^k, and equation (11.5) to show that the end of the zigzag after k steps is at the point with coordinates

$$\mathbf{c} + \frac{l_1}{2\sin\frac{1}{2}\theta_1}\left(\sin\left(k-\frac{1}{2}\right)\theta_1, -\cos\left(k-\frac{1}{2}\right)\theta_1\right)$$

$$+ \frac{l_2}{2\sin\frac{1}{2}\theta_2}\left(\sin\left(k-\frac{1}{2}\right)\theta_2, -\cos\left(k-\frac{1}{2}\right)\theta_2\right). \quad (11.7)$$

(This time you will need the well-known formulae for $\sin(a+b)$ and $\cos(a+b)$.) Of course it is possible to incorporate the formulae obtained above for $\mathbf{c} = \mathbf{c}_1 + \mathbf{c}_2$ but for most purposes it is better to use equation (11.7) which gives the position relative to the 'centre' $\mathbf{c}$.

The size formula
We need to know how far away the point given by equation (11.7) can get from the centre $\mathbf{c}$. Placing $\mathbf{c}$ at the centre of the screen we can then arrange to get all of the zigzag on the screen. Show that, removing the initial term $\mathbf{c}$ from the expression (11.7), the remaining vector has length at most

$$d = \left|\frac{l_1}{2\sin\frac{1}{2}\theta_1}\right| + \left|\frac{l_2}{2\sin\frac{1}{2}\theta_2}\right|. \quad (11.8)$$

Write the vector $\mathbf{c}$ as (a, b). Thus with $\mathbf{c}$ at the centre of the screen we want to make the graphics area a square of side $2d$. This amounts to

$$xl = a - d, \quad xu = a + d, \quad yl = b - d, \quad yu = b + d.$$

Incorporate these into zz2.m to give zz3.m: zigzag program mark 3, which will automatically scale the picture to fit neatly into the screen. Test with several examples, giving details of the ones you test and how well they fit. Remember that in the original setup, $l_1 = 100$.

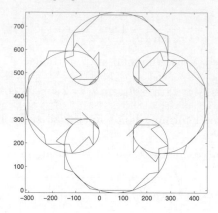

Fig. 11.3. A zigzag (with $l_1 = 100$, $l_2 = 40$, $\theta_1 = 45$, $\theta_2 = 9$), and the associated spirograph which passes through the ends of all the segments of length l_2.

The spirograph

The object here is to draw a 'spirograph' curve which goes through all the positions which the zigzag has reached after k steps, for $k = 1, 2, 3, \ldots$. In fact the curve is given precisely by equation (11.7), except that instead of k taking only integer values $1, 2, \ldots, s$, where s is the number of steps (given by equation (11.6)), it takes on all real values from 0 to s. Thus dividing the interval $[0, s]$ into say 1000 parts we will set up an array of values of k:

```
n=1:1000;
k=s*n/1000;
```

Thus k goes in 1000 steps from $s/1000$ to s. Plot the curve given by equation (11.7) in white after the zigzag has been drawn, calling the resulting M-file `zz4.m`. Give examples. Figure 11.3 shows the case $l_1 = 100$, $l_2 = 40$, $\theta_1 = 45$, $\theta_2 = 9$. Can the spirograph close up before the zigzag does?

The special case $\theta_1 = -\theta_2$

Show that, apart from the initial **c** term in equation (11.7), the remaining terms become, when $\theta_1 = -\theta_2 = \theta$ say,

$$\frac{1}{2\sin\frac{1}{2}\theta}\left((l_1 + l_2)\sin\left(k - \frac{1}{2}\right)\theta, \;\; (-l_1 + l_2)\cos\left(k - \frac{1}{2}\right)\theta\right).$$

Writing this as $(X(k), Y(k))$, where k, as above, is now a real number between 1 and s, show that (X, Y) describes an ellipse with horizontal and vertical axis lengths

$$\left| \frac{l_1 + l_2}{\sin \frac{1}{2} \theta} \right|, \quad \left| \frac{l_1 - l_2}{\sin \frac{1}{2} \theta} \right|$$

respectively. (Note that this ellipse has centre at $\mathbf{c}$, not at the origin, since we ignored the $\mathbf{c}$ in equation (11.7).) Test this out with some values of θ, drawing both the zigzag and the spirograph. How do you make an ellipse with horizontal axis *shorter* than its vertical axis? When is the ellipse a circle?

Spirographs with cusps

The form of equation (11.7) shows that the spirograph curve is actually an epicycloid (compare Chapter 10, Investigation A) obtained by rolling a circle on another circle and tracing the path of a point P rigidly attached to the latter circle. In fact it can be shown that the fixed circle has radius a, the rolling circle has radius b and the point P is at a distance d from the centre of the rolling circle (see Figure 11.4 and note that a, b, d have nothing to do with their previous uses in this project!), where

$$a = \frac{(\theta_1 - \theta_2) l_1}{2 \theta_2 \sin \frac{1}{2} \theta_1}, \quad b = \frac{\theta_1 l_1}{2 \theta_2 \sin \frac{1}{2} \theta_1}, \quad d = \frac{l_2}{2 \sin \frac{1}{2} \theta_2}.$$

(You need not check these.) A case of special interest is $b = d$ which means that the point P is on the *circumference* of the rolling circle. This always produces *cusps* on the spirograph. Show that this amounts to

$$l_2 = \frac{l_1 \theta_1 \sin \frac{1}{2} \theta_2}{\theta_2 \sin \frac{1}{2} \theta_1}.$$

Amend `zz4.m` to `zz5.m` which chooses this automatically for l_2 once θ_1, θ_2 are given (remember $l_1 = 100$).

Investigate cases where there are cusps on the spirograph. For example, what determines the *number* of cusps?

B Fast curves

Aims of the project

We shall examine various candidates for a 'fast curve': taking a smooth wire in the shape of such a curve, we want to minimise the time taken by a bead to slide down the wire. The 'fastest curve' is known to have the shape of a cycloid (see, for example, [17]) but here we shall also use

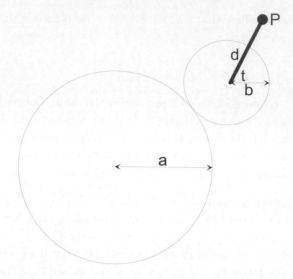

Fig. 11.4. Generation of an epicycloid (spirograph) by a circle rolling outside another circle.

numerical integration techniques to compare the 'speed' of various simple curves such as parabolas and broken lines. There is more information on the same topic in [9].

Mathematical ideas used
This project is about minimisation: finding shortest times of descent. These are expressed by integrals. Some differentiation is involved, though most of the minimisation is done numerically. L'Hôpital's rule is used to evaluate a limit. Various curves such as parabolas and cycloids are used as examples.

MATLAB techniques used
Integration via the `quad8` numerical integration package is used, minimisation of a function of two variables using `fmins` and solution of equations using `fsolve`. There is some use of global variables in the given M-files.

Introduction and formulae for reference (no friction case)
This project is about a bead sliding down a smooth wire, lying in a vertical plane, between two points not in the same vertical line. The object is to investigate the time taken and to compare the times for different shaped curves. There is a famous result, dating back to the

seventeenth century, which says that the shortest time of descent is achieved by a curve called a 'cycloid'. We shall meet the cycloid curve later. The proof that this is the 'curve of shortest descent time' or 'brachistochrone' will not be given here; it is not hard, but uses the first steps in a subject called the calculus of variations. See, for example, [17]. The proof assumes that the wire is smooth: no friction forces act. When there is friction a slightly different result holds; there is a little work on this case at the end of the project.

Consider a curve joining the points $(0, c)$ and $(d, 0)$ in the plane, as in Figure 11.5, where $c = 2, d = 4$. If a bead slides under gravity but without friction or air resistance down this curve, starting at rest at the point $(0, c)$, then energy is conserved. This implies that the kinetic energy at any moment is equal to the potential energy lost up to that moment, so that if v is the speed at time t and y is the y-coordinate at time t, then

$$\frac{1}{2}mv^2 = mg(c - y), \quad \text{i.e.} \quad v = \sqrt{2g(c - y)}, \tag{11.9}$$

where g is the acceleration due to gravity and m is the mass of the bead. Now $v = ds/dt$, where s is the arc-length along the curve, so the time t can be found by integration. Of course the value of g will depend on what units are being used. In order to make the times of descent of a reasonable size (round about 10), the value of g used in the M-files is set at 0.1.

There are various forms of the resulting formula for the total time of descent from $(0, c)$ to $(d, 0)$, depending on whether we integrate with respect to x or whether both x and y are functions of another variable θ say, as will be the case for the cycloid. We state these forms together here; you can assume them during the project. Note that most of the integrations are performed numerically using a MATLAB routine called quad8. At the end of this project there is a note about evaluating the integrals numerically.

The time of travel from $(0, c)$ to $(d, 0)$, where it is assumed that $c \geq 0, d > 0$ and that x increases steadily along the curve, is

$$\text{(i)} \quad \frac{1}{\sqrt{2g}} \int_0^d \sqrt{\frac{1 + y'^2}{c - y}} \, dx, \quad \text{(ii)} \quad \frac{1}{\sqrt{2g}} \int_{\theta_0}^{\theta_1} \sqrt{\frac{x'^2 + y'^2}{c - y}} \, d\theta \tag{11.10}$$

where, in (i), y is a function of x and $'$ means d/dx, while, in (ii), x, y are both functions of θ, $'$ means $d/d\theta$ and θ_0, θ_1 are the values of θ at the points $(0, c)$ and $(d, 0)$ respectively.

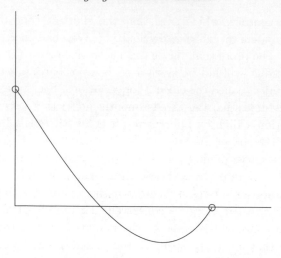

Fig. 11.5. A curve joining the points $(0, 2)$ and $(4, 0)$.

Note that there is a small *problem* in evaluating the integrals above, for, at $x = 0$ in (i) or at $\theta = \theta_0$ in (ii), we have $y = c$ so that the integrand is infinite. Nevertheless, so long as the curve does not go *above* the line $y = c$, the integrals do converge (the time of descent *is* finite!).

MATLAB has a numerical integration package called quad8, but it requires that the integrand is finite over the range of integration. We get round this problem by a completely naive approach†. We use quad8 to integrate over an interval which starts *just after* $x = 0$ (or $\theta = \theta_0$) and use a separate estimate for the remaining tiny piece of the original range of integration. This seems to work reasonably well in practice.

At the end of this project we shall go into some more details on the integrals. For the time being, you can just use the M-files as directed and take it on trust that they give reasonably accurate results!

---

† There are techniques called singularity subtraction techniques which are less naive.

The fastest parabola

Here, we shall take various parabolas between the points $(0, c)$ and $(d, 0)$ and find the times of travel, finding in the process the *fastest parabola*.

(i) Consider a parabola $y = \alpha x^2 + \beta x + c$ which passes through the point $(0, c)$. We shall insist that $\alpha > 0$ so that the parabola is 'convex upwards'. Show that if m is the x-coordinate of the minimum point on the parabola, and if the parabola passes through $(d, 0)$, then

$$\alpha = \frac{c}{d(2m - d)}, \quad \beta = -2\alpha m.$$

Note that in order to fulfil our requirement that $\alpha > 0$ we need $m > \frac{1}{2}d$. We are only interested in that part of the parabola between $x = 0$ and $x = d$. If $\frac{1}{2}d < m < d$ then the bead goes below the x-axis before rising to the point $(d, 0)$ while if $m \geq d$ then it is downhill all the way for the bead.

(ii) Take $c = 2, d = 4$. *Sketch* the parabolas for $m = 2.5, 3, 4$ and 5. (The M-file `paramc.m`, which plots parametric curves, is available if you really need it, but it would be far quicker here to sketch by hand, knowing the general shape of a parabola.) Make a *guess* as to which parabola you think might give the fastest ride for a bead sliding from $(0, 2)$ to $(4, 0)$.

(iii) The M-file `slide1.m` uses the MATLAB package `quad8`, plus a separate approximation for $0 \leq x \leq 0.01$ (see the Introduction above and the note at the end of this project) to evaluate the integral (i) in equation (11.10) in the case of the parabola. It prints out these two times and then the total time of descent is given.

The M-file requires another function file containing the function to be integrated, and this is called `slide1fn.m`. Look carefully at both these M-files since you will need to amend them for other curves later. In particular look at the (rather tiresome) need for global variables in these two M-files. Both the function y and its derivative, called $y1$ in the M-files, are used.

Keeping to $c = 2, d = 4$, find the times of descent for $m = 2.5, 3, 4$ and 5.

Now amend the M-file (calling it `slide2.m`) so that it takes values of m from say $\frac{1}{2}d + 0.1$ to d in steps of 0.1 and finds the time of descent for each one, storing it in a vector called `result`, finally plotting the graph of m against the time. So you will want something like

```
m=d/2+.1:.1:d;
```

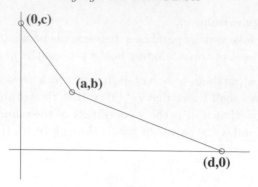

Fig. 11.6. A broken line joining $(0, c)$ and $(d, 0)$, the break point being (a, b).

```
result=zeros(length(m),1);
for i=1:length(m)
.......
result(i)=time;
end % of i=1:length(m) loop
plot(m,result)
```

Note that the `global` statement in `slide1.m` MUST remain outside the `for` loop! There is no need to change `slide1fn.m`.

For $c = 2, d = 4$ print out the graph of m against time of descent and estimate from it the value of m which gives the minimum time of descent. Draw a hand sketch of the corresponding parabola. (Was your guess right?!) For $c = 2, d = 3$ use the same method to find the 'fastest parabola' and sketch it, but do not print out the graph of m against the time.

Broken lines

Suppose we join $(0, c)$ and $(d, 0)$ by two straight segments, joining at the 'break point' (a, b) say. See Figure 11.6. We shall look for the fastest broken line. So here c and d are given, and a and b can be changed.

(iv) Show that the equations of the two segments are

$$y = c + \frac{(b-c)x}{a} \ \ (x \le a), \quad y = \frac{b(x-d)}{a-d} \ \ (x > a).$$

(v) For a straight line descent it is actually possible to calculate the time exactly, by evaluating the first integral in equation (11.10). Consider a line of slope m, between (x_0, y_0) and (x_1, y_1); note that m will generally be negative here. Replacing dx by dy/m in the integral, show that the time of descent from height y_0 to height y_1 is

$$\frac{\sqrt{1+m^2}}{m\sqrt{2g}} \int_{y_0}^{y_1} \frac{1}{\sqrt{c-y}} dy = \frac{2\sqrt{1+m^2}}{m\sqrt{2g}} \left(\sqrt{c-y_0} - \sqrt{c-y_1} \right). \quad (11.11)$$

Taking $c = 2, d = 4$ as before (and $0 < a < d, 0 \le b < c$), show that the time of descent along the broken line is

$$\frac{2}{\sqrt{2g}} \left(\sqrt{\frac{a^2 + (2-b)^2}{2-b}} + \frac{(\sqrt{2} - \sqrt{2-b})\sqrt{(4-a)^2 + b^2}}{b} \right). \quad (11.12)$$

You can use the built-in MATLAB function `fmins` to find the minimum of the function of a and b defined by equation (11.12). You have to make an M-file, say `slide3fn.m`, containing the function to be minimised. This takes the form

```
function realtime=slide3fn(p)
g=.1; c=2; d=4;
a=p(1); b=p(2);
realtime = ;% Formula for the real time of descent,
% as above
```

You then call `fmins` by `fmins('slide3fn',[*,*]')`, where the asterisks are replaced by a sensible guess as to the values of a, b which give the minimum. (Note the ' after the vector, indicating transpose.) Make a sensible guess and find the values giving the minimum time of descent. (Note that if you make a guess of say $a = 1, b = 2.1 > c$ then, as you might expect, `fmins` does not give a reasonable answer.) Draw a sketch of the broken line which achieves the minimum time. Also find the corresponding minimum time. How does this compare with the fastest parabola?

(vi) An interesting case of equation (11.12) is when $b = 0$, which makes the second term now of the form $0/0$. Use L'Hôpital's rule on the expression

$$\frac{\sqrt{2} - \sqrt{2-b}}{b}$$

to show that when $b = 0$ the time of descent is (still with $c = 2, d = 4, a < d$),

$$\frac{1}{2\sqrt{g}} \left(2\sqrt{a^2 + 4} + 4 - a \right).$$

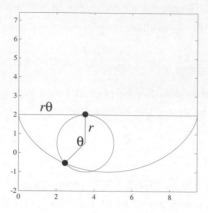

Fig. 11.7. One complete arch of the cycloid starting at (0,2), generated by rolling a circle of radius $r = 1.5$ underneath the line $y = 2$. The cycloid is parametrised by θ.

Show, using calculus, that the value of a between 0 and 4 which makes this a minimum is $2/\sqrt{3}$.

The cycloid

A cycloid curve is obtained by rolling a circle along a straight line and tracing the path of a point attached to the circumference of the circle. In our case we roll a circle of radius r 'under' the straight line $y = c$. See Figure 11.7. The cycloid is parametrised by the angle θ in the figure; the coordinates of the moving point P are

$$x = r(\theta - \sin\theta), \quad y = c - r + r\cos\theta. \qquad (11.13)$$

Figure 11.7 shows a complete 'arch' of a cycloid, for $0 \le \theta \le 2\pi$. Note that $\theta = 0$ gives $(x, y) = (0, c) = (0, 2)$ in the figure.

As before we take c and d as given. We want the cycloid to pass through the point $(d, 0)$; in order for this to happen we need, for some $r > 0$ and $0 < \theta \le 2\pi$,

$$r\sin\theta - r\theta + d = 0, \quad r\cos\theta - r + c = 0. \qquad (11.14)$$

Since these are two equations for two unknowns r, θ it is reasonable that they might have a solution, maybe even a unique solution. Since they are rather unpleasant equations, we have to solve them using one of MATLAB's equation solvers, e.g. `fsolve`. (If by any chance your version of MATLAB lacks `fsolve` then you could try using one of the custom written M-files from Chapter 15 called `full_new.m` and `gauss_ja.m`.)

(vii) Taking $c = 2, d = 4$, find the solution of equation (11.14) as follows. Create a function file say `cycfun.m` of the following form

```
function q=cycfun(p)
c=2; d=4;
r=p(1); theta=p(2);
q=zeros(2,1);
q(1)=r*sin(theta)-r*theta+d;
q(2)=r*cos(theta)-r+c;
```

You can then solve the equations (11.14) with $c = 2, d = 4$ by

`fsolve('cycfun',[r0, theta0]')`

where `r0` and `theta0` are a first guess as to the solution for r and θ. Don't forget the ' after the vector, indicating transpose. Make a reasonable guess and find the resulting solution. (*Hint*: Try $r = 1$ for the radius.)

Denote the solution for θ by θ_1 (referred to as `theta1` in the M-files). Sketch the cycloid over the range from $\theta = 0$ to $\theta = \theta_1$, i.e. between the points $(0, 2)$ and $(4, 0)$.

The M-file `slide4.m` uses the second formula of (11.10) to calculate the time of descent when a curve is parametrised by θ. The range of values of θ is from 0 to θ_1. This M-file uses a function `slide4fn.m` in the same way that `slide1.m` uses the function `slide1fn.m`.

What is the time of descent for the cycloid? There is a famous theorem, first proved in the seventeenth century by Newton, Leibniz, Bernoulli and others, that the cycloid is the curve which gives the fastest descent. How much faster is it (as a percentage) than the fastest broken line?

(viii) Notice that (in the present friction-free case) the cycloid is still best when $c = 0$, i.e. when the start and end are on the same horizontal level. In this case the bead, on arriving at $(d, 0)$, has zero speed. When $c = 0, d = 4$ what is the time taken along the cycloid? Beware that you might need to take θ_1 just a little less than the true value to avoid an infinite time being produced by `slide4.m`. Alternatively, with a little cunning you can find the time it takes to travel half-way along the arch of this cycloid and double the answer.

The square root curve

Find the value of α so that the curve

$$y = \alpha\sqrt{x} + c,$$

which passes through $(0, c)$, also passes through $(d, 0)$. Find the time of travel between these points along this curve in the case $c = 2, d = 4$ and also sketch the curve in this case. You can adapt `slide1.m` here, calling the result `slide5.m`, remembering to produce also an M-file `slide5fn.m` which contains the function y and its derivative $y1$, as in `slide1fn.m`. How does this time compare with that for the cycloid?

The integrals (optional reading)

Using the formula (11.9) in the form

$$\frac{ds}{dt} = \sqrt{2g(c - y)},$$

and the standard formulae

$$\frac{ds}{dt} = \frac{ds}{dx}\frac{dx}{dt} = \sqrt{1 + y'^2}\frac{dx}{dt}$$

(where $y' = dy/dx$), prove formula (11.10)(i).

As was pointed out in the introduction to this chapter, MATLAB's numerical integrator assumes that the function being integrated is finite over the range of integration, so in this project the integrator is used in equation (11.10)(i) over the range $0.01 \leq x \leq d$ and a separate estimate is obtained over the initial small interval $0 \leq x \leq 0.01$. In fact, to obtain the estimate, we shall assume that $y' = dy/dx$ is *constant* for $0 \leq x \leq 0.01$. It can be shown that the integral between these limits is then given by the formula of (11.11), namely

$$-\frac{\sqrt{2(1 + y'^2)(c - y)}}{y'\sqrt{g}},$$

(note that y' will be negative).

In the approximations used in this project, the values of y and y' are taken at the endpoint $x = 0.01$.

Friction (optional extra)

It is interesting to make the above considerations more 'realistic' by introducing friction. This is surprisingly easy to do, since the only effect is to replace the formula (11.9) by the slightly more complicated formula

$$v = \sqrt{2g(c - y - \mu x)},$$

where μ is the coefficient of (kinetic) friction. This means that the bead is acted on by gravity downwards, the normal reaction of the wire, and friction, which is μ times the normal reaction. The equation for v then follows from Newton's second law of motion (you need not check this unless you want to!).

As a result, the formula (11.10)(i) for the time of travel becomes

$$\frac{1}{\sqrt{2g}} \int_0^d \sqrt{\frac{1+y'^2}{c-y-\mu x}}\, dx.$$

Adapt `slide1.m` and `slide1fn.m` to find the fastest parabola when $\mu = 0.3$. Call the M-files `slide6.m` and `slide6fn.m` . Beware that taking values of μ larger than about 0.5 may result in 'infinite' times of travel along some parabolas. Why is this?

12

Sequences of Real Numbers

This chapter is about sequences of real numbers, and we begin with an introduction to special classes of sequences. There follow three investigations which explore some more properties. The first (A) is on Möbius sequences, which are described below, while the second (B) is entirely on quadratic sequences. The third one (C) is mostly on quadratic and exponential sequences but there is some mention of Möbius sequences again. Check with each investigation to see which parts of the preliminary material (§§12.1–12.3) are required.

12.1 Möbius sequences

We shall consider several sequences of the form

$$x_{n+1} = \frac{ax_n + b}{cx_n + d}, \ n = 0, 1, 2, 3, \ldots,$$

where a, b, c, d are real numbers. Once x_0, also real, is given, the whole sequence is determined and consists of real numbers. It is called a Möbius sequence. There are two M-files available. The first is mobius.m which calls for the numbers a, b, c, d, x_0 and the number of iterations to be performed, that is, how many terms of the sequence are to be calculated. It then plots a bar-chart of the values.

The second M-file is mobius1.m which dispays the values of the successive x_i as a column vector, using 'long format' for greater accuracy.

(i) $(a, b, c, d) = (1, 2, 1, 1), x_0 = 3$. It is fairly clear from mobius.m and mobius1.m (use 20 iterations) that this sequence is *convergent*. The limit also looks suspiciously familiar. Running with other values of x_0 produces the same result. Note that some values of x_0 produce ∞ on the way, as with $x_0 = -1.5$ for example. (Here, $x_2 = \infty$.) However, this is not a serious problem since the formula suggests that if $x_n = \infty$ then x_{n+1} should be simply a/c. In fact the M-files take account of this and do not crash if an infinite value is obtained. (In practice it is more usually a *very large* value, positive or negative.)

The M-file `mobius.m` is reproduced at the end of this section to show you how the 'infinity' problem is overcome.

(ii) $(a, b, c, d) = (1, 1, -2, 1), x_0 = 3$. Using `mobius.m` and 500 iterations seems not to produce a limit. Using `mobius1.m` and 20 iterations it certainly seems that the numbers are jumping around a lot. Such a sequence could be called 'wildly divergent' or 'oscillatory' or 'chaotic'.

(iii) $(a, b, c, d) = (1, 1, -1, 1), x_0 = 3$. This gives a repeating cycle of four values $3, -2, -\frac{1}{3}, \frac{1}{2}$, which can be seen by either of the two M-files. Such a sequence is said to be *periodic with period four*.

Clearly a lot of different behaviours are possible. Note that, in the last two above, we can argue that *if* $x_n \to l$, *then* $x_{n+1} \to l$ too, so that, respectively,

$$l = \frac{l+1}{-2l+1}; \text{ thus } l^2 = -\frac{1}{2},$$

$$l = \frac{l+1}{-l+1}; \text{ thus } l^2 = -1.$$

In both cases this is a contradiction (remember we are working with real numbers). Hence it is certain that neither of these two sequences can have a single limit l, but the modes of divergence are very different. Note that an apparently wildly divergent sequence might in fact be periodic with a very large period, and this might be obscured by numerical inaccuracies in computation. There is no easy answer to this—and we shall not look for an answer here!

Applying the above method to the first example shows that *if* the sequence has a limit *then* that limit must be $\pm\sqrt{2}$. It is not quite so easy to prove that it *does* have a limit, but this can be done. The 'cobweb diagram' we shall meet shortly makes this very plausible indeed—in fact it can be made into a proof. It is slightly mysterious that $l = -\sqrt{2}$ is *never* the limit unless you actually start with $x_0 = -\sqrt{2}$.

Here is the M-file `mobius.m`. Note that x_n can be '$\pm\infty$'—this happens when $x_{n-1} = -d/c$. But it is still possible to calculate x_{n+1}, namely $x_{n+1} = a/c$. Note that 'infinity' here just means a very large number, which in MATLAB is referred to as `Inf`.

```
% M-file to plot Mobius sequences as bar chart.
% If x is the current term then the next one
% is (ax+b)/(cx+d).
% The constants a,b,c,d are input by the user.
% You are also asked for the number of iterations
% to be performed and the starting value x0.

a=input('Type a ');
b=input('Type b ');
c=input('Type c ');
d=input('Type d ');

x0=input('Type x0 ');
n=input('Type the number of times to be iterated ');

x=[x0];

for j=1:n
  if x(j)==Inf | x(j)==-Inf
  x(j+1)=a/c;
  else
  x(j+1)=(a*x(j)+b)/(c*x(j)+d);
  end
end;

bar(x)
```

12.2 Cobweb diagrams

These are a diagrammatic way of following sequences under iteration. Given a function $y = f(x)$, we start with x_0 and find in succession $x_1 = f(x_0)$, $x_2 = f(x_1)$, etc. Start with the graph $y = f(x)$ and add the line $y = x$; then proceed as follows. From the point $(x_0, 0)$ on the x-axis a vertical line is drawn to meet the graph at (x_0, x_1), then a horizontal line to meet the line $y = x$ at (x_1, x_1). The process is then repeated: a vertical line to meet the graph at (x_1, x_2), then a horizontal line to meet $y = x$ at (x_2, x_2), and so on. The successive vertical lines have x-coordinates $x_1, x_2, \ldots$. Thus the progress of the sequence can be followed.

Figure 12.1 shows two examples of cobweb diagrams, one in fact for a Möbius sequence and one for a quadratic sequence (these are discussed in Project B below). Possibly the second one gives more of an idea of why the name 'cobweb' is used!

Möbius sequence cobwebs

The M-file `cobm.m` plots cobweb diagrams for Möbius sequences, which use the function $f(x) = (ax+b)/(cx+d)$. It requires input of a, b, c, d, x_0 and the upper and lower limits of x and y on the diagram. Iteration is done 30 times, or until it seems that convergence is certain. The sequence of values of x is printed out at the end.

Note that, when there is a vertical asymptote to the graph of the function f, this will be replaced by a vertical line joining two points of the graph. Here are some suitable values:

1 In Example (i) above, lower limits of x and y equal to -3 and upper limits of x and y equal to 3, $x_0 = 2$ or -2. If you want to see the action in close-up then you could take say $1.3 \leq x \leq 1.5$ and the same for y.

2 In Example (ii) above, lower limits of x and y equal to -10 and upper limits of x and y equal to 10, $x_0 = 2$.

3 In Example (iii) above, lower limits of x and y equal to -5 and upper limits of x and y equal to 5, $x_0 = 3$.

It is fairly clear in each case what is happening.

The points where the graph of f meets the line $y = x$ are *fixed points* of the function y, that is, writing $y = f(x)$, they satisfy $f(x) = x$. It is these points, if any, which are limits of the sequence.

12.3 Möbius functions and powers of matrices

Let

$$A = \begin{pmatrix} a & b \\ c & d \end{pmatrix}, \qquad x_{n+1} = \frac{ax_n + b}{cx_n + d}.$$

Clearly there is going to be a close connection between the matrix A and the Möbius sequence. Here is one of many ways of expressing this connection.

Theorem 1 (i) *There is a real number α (depending on n) such that*

$$A \begin{pmatrix} x_n \\ 1 \end{pmatrix} = \alpha \begin{pmatrix} x_{n+1} \\ 1 \end{pmatrix}.$$

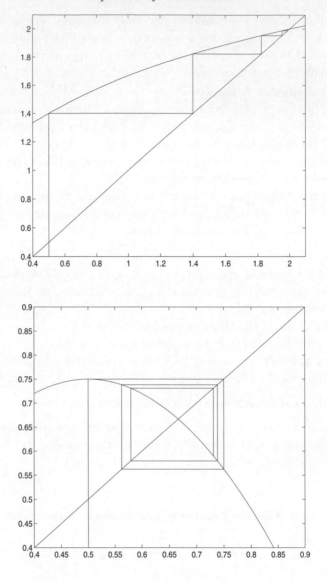

Fig. 12.1. Two cobweb diagrams, the upper one corresponding to a Möbius sequence and the lower one to a quadratic sequence.

In other words,

$$\text{if } A \begin{pmatrix} x_n \\ 1 \end{pmatrix} = \begin{pmatrix} w_1 \\ w_2 \end{pmatrix} \text{ then } x_{n+1} = \frac{w_1}{w_2}.$$

(ii) *For any integer $k \geq 1$ there is a real number β (depending on k) such that*

$$A^k \begin{pmatrix} x_0 \\ 1 \end{pmatrix} = \beta \begin{pmatrix} x_k \\ 1 \end{pmatrix}.$$

In other words,

$$\text{if } A^k \begin{pmatrix} x_0 \\ 1 \end{pmatrix} = \begin{pmatrix} w_1 \\ w_2 \end{pmatrix} \text{ then } x_k = \frac{w_1}{w_2}.$$

Example 1
Let

$$A = \begin{pmatrix} 1 & 2 \\ 1 & 1 \end{pmatrix}, \quad x_0 = 3.$$

Then with $\mathbf{v} = (3,1)^\top$, we have $A\mathbf{v} = (5,4)^\top$ and the α in (i) of Theorem 1 is 4, while $x_1 = \frac{5}{4}$. Taking $k = 4$ in (ii), $A^4\mathbf{v} = (75,53)^\top$ and $x_4 = \frac{75}{53} = 1.4151$; the value of β is 53.

Example 2
If you type

```
format long
A=[1 2; 1 1];
v=[2 1]';
w=A^20*v;
w(1)/w(2)
```

then you should get the answer 1.41421356237310, which is the same answer as you get by typing `sqrt(2)` . By (ii) of the Theorem 1, the number obtained is x_{20} when $x_0 = 2$.

The proof of Theorem 1 is straightforward: (i) is proved by multiplying out the two sides and taking $\alpha = cx_n + d$. Then (ii) is proved by induction on k. The case $k = 1$ is just (i) with $n = 0$. Assuming (ii) for k, we proceed as follows:

$$A^{k+1} \begin{pmatrix} x_0 \\ 1 \end{pmatrix} = A\beta \begin{pmatrix} x_k \\ 1 \end{pmatrix} = \beta A \begin{pmatrix} x_k \\ 1 \end{pmatrix} = \beta\alpha \begin{pmatrix} x_{k+1} \\ 1 \end{pmatrix}.$$

The first $=$ uses the induction hypothesis, the second $=$ uses the fact that β is just a number and the third $=$ uses (i) with n replaced by k. This completes the proof of (ii) by induction. □

If A^k is a *scalar* matrix—that is, of the form λI, where I is the identity

2×2 matrix—then we have

$$A^k \begin{pmatrix} x_0 \\ 1 \end{pmatrix} = \lambda \begin{pmatrix} x_0 \\ 1 \end{pmatrix},$$

so that by (ii) of the theorem $x_k = x_0$, i.e. the values of the Möbius sequence have recycled back to x_0. So periodic sequences can be detected by calculating powers of the corresponding matrix.

We can also work 'backwards': for example, let us work out the starting value x_0 which, for the sequence

$$x_{n+1} = \frac{3x_n + 5}{x_n + 2},$$

gives $x_6 = -2$. Note that when x_0 is chosen in this way, the next term x_7 will be 'infinite', since the denominator $x_6 + 2$ vanishes. We want

$$A^6 \begin{pmatrix} x_0 \\ 1 \end{pmatrix} = \beta \begin{pmatrix} -2 \\ 1 \end{pmatrix}$$

for some β. This gives

$$\frac{1}{\beta} \begin{pmatrix} x_0 \\ 1 \end{pmatrix} = A^{-6} \begin{pmatrix} -2 \\ 1 \end{pmatrix} = \begin{pmatrix} -22658 \\ 12649 \end{pmatrix},$$

where you can verify the last equality by typing `A^(-6)*[-2 1]'` having first input the matrix

$$A = \begin{pmatrix} 3 & 5 \\ 1 & 2 \end{pmatrix}.$$

(You may find that -22658 actually appears as something like -22657.999 of course!) Thus $x_0 = -22658/12649 = -1.7913$. If you use `mobius1.m` to find x_7 for this sequence (input x_0 in the above fractional form) you will find that it gives a very large value. Nevertheless x_8 is given correctly as 3.

There is a highly significant fact about the *direction* of the vector $\mathbf{w} = A^k \mathbf{v}$ for large values of k. Going back to the matrix A of Example 2 above, try typing

`[X D]=eig(A)`

You will find that one of the eigenvectors of A is parallel to the vector $\mathbf{w}$ obtained in Example 2. For example, you can type

`x1=X(:,1)`
`y=x1./w`

to pull out the first eigenvector of A and then compare it with $\mathbf{w}$ in direction. The two entries in the vector $\mathbf{y}$ should be approximately the same. Note that the eigenvector used here is in fact the one corresponding to the *larger* eigenvalue.

With matrices we can equally well consider 3×3, but note that this does not have an obvious analogue in terms of sequences. Try

```
A=[1   2   3;-1   2   -3;1   -2   -3]
```

and (almost) any starting vector $\mathbf{v}$. This matrix has *one* real eigenvector, $(-0.5433, 0.2849, 0.7897)$, as you can discover by typing $[\texttt{X},\texttt{D}]\ =\ \texttt{eig(A)}$. This has the same direction as $A^n\mathbf{v}$ for large values of n.

Note on two M-files
It can happen that powers of a matrix A have entries which become very large indeed, and working out $A^n\mathbf{v}$ for a large n can be hazardous. There are two M-files available which overcome this problem when we only want to know the *direction* of $A^n\mathbf{v}$. They work out $A\mathbf{v}, A^2\mathbf{v}, A^3\mathbf{v}, \ldots$ but after each step rescale the answer to be a unit vector. They are called $\texttt{matit2.m}$ for 2×2 matrices and $\texttt{matit3.m}$ for 3×3 matrices. Both require previous input of the matrix A but not of the vector $\mathbf{v}$. Look at them if you are curious.

For those who like to know the theory, we have the following.

Theorem 2 *If $A^n\mathbf{v}$ does have a limiting direction, then this limit must be an eigenvector of the matrix A.*

Proof If we suppose $A^n\mathbf{v}/\|A^n\mathbf{v}\| \to \mathbf{w}$ for some $\mathbf{w} \neq 0$ as $n \to \infty$, then $A(A^n\mathbf{v}/\|A^n\mathbf{v}\|) \to A\mathbf{w}$, so

$$\frac{A^{n+1}\mathbf{v}}{\|A^{n+1}\mathbf{v}\|} \cdot \frac{\|A^{n+1}\mathbf{v}\|}{\|A^n\mathbf{v}\|} \to A\mathbf{w}.$$

But, as $n \to \infty$ the first fraction here tends to $\mathbf{w}$ and the second to some real number α, so we obtain $A\mathbf{w} = \alpha\mathbf{w}$, which means that $\mathbf{w}$ is an eigenvector of A. Of course this does not prove that $A^n\mathbf{v}$ *does* have a limiting direction equal to an eigenvector, only that eigenvectors are the *only possible* limiting directions. $\square$

A Investigation on Möbius sequences

Aims of the project
We shall explore Möbius sequences both experimentally and theoretically. You should read through §§12.1–12.3 first.

Table 12.1. *Values to try for the Möbius sequence.*

a	b	c	d
2	1	1	−2
3	2	4	3
3	2	1	2
1	3	−2	1
1	$-\sqrt{3}$	$\sqrt{3}$	1
$\sqrt{3}$	−1	1	$\sqrt{3}$

Mathematical ideas used

Matrices, sequences, mathematical induction and limits of sequences. There is some calculation of derivatives in the theory part of the project. You should read through §§12.1–12.3 first.

MATLAB techniques used

Only pre-written M-files are used here to iterate Möbius sequences, to find powers of matrices and to plot cobweb diagrams.

Möbius functions and matrices

(i) For the values of a, b, c, d given by the Table 12.1 decide using the M-files `mobius.m` and/or `mobius1.m` whether the corresponding Möbius sequence is convergent or divergent, and in the latter case whether it appears to be 'wildly divergent' ('chaotic') or appears to be periodic (cycling round in a loop). Write down your results, specifying your initial value(s) x_0 and the numbers of iterations you used.

Also use the M-file `cobm.m` to plot cobweb diagrams. To iterate you use <Enter>. Make rough sketches of these on paper in the cases where you assert the sequence is convergent or periodic, specifying your range of values of x and y as well as starting value(s) x_0.

Note: Use the expression `sqrt(3)` when prompted for an input. MATLAB will substitute the numerical value for you.

(ii) Take the same examples as in the previous exercise, associating (a, b, c, d) with the matrix

$$\begin{pmatrix} a & b \\ c & d \end{pmatrix}.$$

Use the matrix multiplication of MATLAB to verify that, for the *periodic* sequences, the nth power of the matrix ($n =$ the length of the period) is a *scalar* matrix. This means a matrix of the form

$$\lambda I = \begin{pmatrix} \lambda & 0 \\ 0 & \lambda \end{pmatrix},$$

where I is the 2×2 identity matrix. Write down the value of λ in each case. Also verify that, for $(a, b, c, d) = (1, 1, -1, 1)$, the fourth power of the matrix is a scalar matrix, and write down the value of λ.

(iii) What is the connection between $A^4 = \lambda I$ in the last part of the previous exercise and the fact that the corresponding Möbius sequence is periodic with period 4? (See §12.3.)

(iv) For which of the examples in Exercise (i) does the M-file `matit2.m` give a limiting direction for the iterated product $A^n \mathbf{v}$? State the value(s) of the vector $\mathbf{v}$ which you used and the limiting direction obtained. Verify that this is an eigenvector of A by calculating the eigenvectors. State both eigenvectors in these cases.

(v) For $(a, b, c, d) = (1, 2, 1, 1)$, find the inverse of the corresponding matrix (e.g. by using `inv(A)`). Hence find (exactly) a starting value x_0 for which $x_5 = \infty$. Is there any possibility of finding x_0 such that $x_{50} = \infty$?

(vi) *This part has nothing to do with Möbius functions but involves powers of 3×3 matrices.* Compare §12.3. A population of bugs is of three kinds: black, red and green. At time $t = 0$ there are 10 black ones but no red or green ones. At time $t = 1$, the following three things happen simultaneously:

(a) each black bug splits into a black bug, 2 red bugs and a green bug;
(b) each red bug splits into 1 black bug and 2 green bugs;
(c) each green bug splits into 3 red bugs.

The same happens at times $t = 2, t = 3$, etc. The bugs in this problem are immortal (this is obviously pure mathematics).

How can you use matrix iteration to discover, for large values of t, what percentages of the population are black, red and green? Does the initial population (at $t = 0$) make any difference? Is there a connection with one of the eigenvectors of a matrix? *Hint*: Consider the matrix

$$\begin{pmatrix} 1 & 1 & 0 \\ 2 & 0 & 3 \\ 1 & 2 & 0 \end{pmatrix}.$$

Theory of Möbius sequences

Here you have the opportunity to work out some of the theory of these sequences. Illustrate the theory with suitable examples, using the M-files.

(vii) Let $f(x) = (ax + b)/(cx + d)$, where $c \neq 0$ and $ad - bc \neq 0$. The *fixed points* of f are numbers x such that $f(x) = x$. Show that if $(a - d)^2 + 4bc > 0$ then there are two such *real* fixed points α and β, say.

(viii) Remembering $x_{n+1} = f(x_n), f(\alpha) = \alpha, f(\beta) = \beta$, show that

$$\frac{x_{n+1} - \alpha}{x_{n+1} - \beta} = \left(\frac{c\beta + d}{c\alpha + d}\right) \left(\frac{x_n - \alpha}{x_n - \beta}\right). \tag{12.1}$$

Hint: First show, using $\alpha = (a\alpha + b)/(c\alpha + d)$, that

$$x_{n+1} - \alpha = \frac{(ad - bc)(x_n - \alpha)}{(cx_n + d)(c\alpha + d)},$$

and a similar formula for $x_{n+1} - \beta$.

Now deduce, by repeated application of equation (12.1) for $n - 1, n - 2, \ldots$, that

$$\frac{x_n - \alpha}{x_n - \beta} = \left(\frac{c\beta + d}{c\alpha + d}\right)^n \left(\frac{x_0 - \alpha}{x_0 - \beta}\right).$$

(ix) Show that $f'(\alpha) = (c\beta + d)/(c\alpha + d)$, and $f'(\beta) = (c\alpha + d)/(c\beta + d)$. *Hint*: It is easy to check that $f'(x) = (ad - bc)/(cx + d)^2$. Remembering that α and β are the roots of $cx^2 - (a - d)x - b = 0$, deduce that $(c\alpha + d)(c\beta + d) = ad - bc$.

(x) Deduce that if $\mid f'(\alpha) \mid < 1$, then $x_n \to \alpha$ as $n \to \infty$.

(xi) Show that if $(a - d)^2 + 4bc < 0$, then the fixed points are not real. Why does this imply that the sequence $\{x_n\}$ cannot have a limit?

B Attracting cycles

Aims of the project

This project is about sequences formed by iteration of a quadratic (degree 2) function. These sequences do not generally have a unique limit, but have 'many limits' in the sense that successive values of the sequence come close to a collection of different numbers. We shall investigate these 'attracting cycles' both mathematically and experimentally. More details of the mathematics involved here can be found in [2]. There is no need to read the material on Möbius sequences in order to do this project.

Mathematical ideas used

We use sequences of real numbers defined by a quadratic formula. There is some manipulation of polynomials and curve sketching involved in the calculations.

MATLAB techniques used

The project uses existing M-files for analysing quadratic sequences, and also to find the roots of a polynomial equation.

We shall not go much into the theory of attracting cycles; the point of this investigation is to find some examples. We write $f(x) = \lambda x(1 - x)$ and $x_{n+1} = f(x_n)$, $n = 0, 1, 2, \ldots$, to define the *quadratic sequence* $x_0, x_1, x_2, \ldots$. Once λ is given, this sequence is determined by the first term x_0.

Write f^p for the pth iterate of f, that is, $f^2(x) = f(f(x)), f^3(x) = f(f(f(x)))$, etc. Thus $x_p = f^p(x_0)$. For a given value of p, we are particularly interested in finding values of λ such that

$$f^p\left(\frac{1}{2}\right) = \frac{1}{2}. \tag{12.2}$$

This is because of the following result: If λ is such a value, then, for some q which is a *factor* of p, the numbers $\frac{1}{2}, f^i(\frac{1}{2})$, for $i = 1, \ldots, q - 1$ form an 'attracting q-cycle'. This means that if you start from x_0 in the interval $(0, 1)$ and iterate f, then the values $f^p(x_0)$, $p = 1, 2, 3, \ldots$, come close in succession to these q numbers.

(i) For $p = 2$ show that $f^2(x) = \lambda^2 x(1 - x)(1 - \lambda x + \lambda x^2)$ and hence show that equation (12.2) becomes $\lambda^3 - 4\lambda^2 + 8 = 0$. Put the coefficients into a vector v (starting with the leading term) and use `roots(v)` to find the roots of this equation. Ignore the negative root. One positive root is 2; for this value check that in fact $f(\frac{1}{2}) = \frac{1}{2}$. For the other root, use `cobq.m` to confirm that, starting with various x_0, the values of the sequence approach *two* numbers. Write down these numbers as given by the M-file.

(ii) The M-file `quadn.m` plots the graph of any iterate of f (for any given λ), and plots the lines $x = \frac{1}{2}, y = \frac{1}{2}$ in green on the same diagram. It allows you to change λ without changing the other settings. Instructions appear on the screen. Note that after a graph is drawn, you will need to press <Enter> to return control to the main MATLAB window. Then you will be prompted to press 0 to stop, or 1 to choose another λ.

By homing in on the point $(\frac{1}{2}, \frac{1}{2})$ you can get an extremely accurate value of λ such that equation (12.2) holds. (Adjust λ until the graph goes through the intersection of the green lines.) Use it for the third iterate f^3, and find the value of λ near to 3.83 such that equation (12.2) holds. You should be able to manage a couple more decimal places at least, with judicious use of the M-file. Having found λ, use cobq.m to check that, starting with various x_0, the sequence approaches the 'attracting 3-cycle' as above. Write down the numbers occurring in this 3-cycle. For each x_0 that you use, state how many iterations were needed before the sequence approached these numbers to four decimal places.

(iii) For $p = 6$ there is a solution of equation (12.2) near to 3.63. Find this solution as nearly as you can, using quadn.m, and use cobq.m to find the set of six numbers (the 'attracting 6-cycle') which the sequence approaches. For $p = 16$ there is a solution near to $\lambda = 3.55$. Does this give an attracting 16-cycle? Give details (using the M-files, of course).

(iv) Show that the (12.2) for $p = 3$ is

$$\lambda^7 - 8\lambda^6 + 16\lambda^5 + 16\lambda^4 - 64\lambda^3 + 128 = 0.$$

Put the coefficients into a vector p (starting with the leading coefficient 1) and find the solutions with roots(p). Ignore negative real roots and complex roots. One root should be 2; check that for this λ, $f(\frac{1}{2}) = \frac{1}{2}$. For the other root, use cobq.m to verify that there really is an attracting 3-cycle, and write down the numbers in it (you should get the same limiting numbers whatever x_0 in $(0, 1)$ you start from).

(v) For $p = 4$ there is a solution near to $\lambda = 3.5$. Find this solution to several more decimal places, using quadn.m, and use it in cobq.m to find the numbers in the resulting attracting 4-cycle.

(vi) Show that the graph $y = f(x)$ crosses the line $y = x$ when $x = 0$ or $x = 1 - \frac{1}{\lambda}$. Deduce that, for $0 < \lambda < 1$, the graph of $y = f(x)$ is entirely under the graph of $y = x$ for $0 < x < 1$. Explain with a (hand-drawn) cobweb diagram how this makes it very plausible that, for any x_0 with $0 < x_0 < 1$, the resulting quadratic sequence is convergent with limit 0.

For $1 < \lambda < 2$ show that the graph $y = f(x)$ crosses the line $y = x$ just once for $0 < x < 1$, namely for $x = 1 - \frac{1}{\lambda}$. Calculate the slope of the graph $y = f(x)$ where this happens, showing that it is positive. Again using a hand-drawn cobweb diagram explain why this makes it plausible

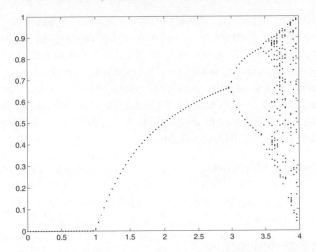

Fig. 12.2. The 'period doubling' diagram. For each value of λ on the horizontal axis, the points above it are the values of x in the resulting attracting cycle. So from λ around 3.5 there are longer and longer attracting cycles.

that, for $0 < x_0 < 1$, the resulting sequence is convergent. What is the limit this time? Confirm your assertion by means of some examples, using the M-files.

(vii) The M-file `perdoub.m` (for 'period doubling'!)   takes 100 values of λ between two chosen limits $l1$ and $l2$. For each one the numbers $x_1, \ldots, x_{100}$ are calculated (with $x_0 = \frac{1}{2}$) and *just the terms from the* 75th *on* are plotted on the vertical line above the value λ. So the picture has axes λ, x, where $l1 \leq \lambda \leq l2$ and $0 \leq x \leq 1$. Figure 12.2 shows the picture for $1 \leq l \leq 4$.

   Try running it for say $l1 = 0, l2 = 2$. How does the resulting picture confirm what is shown in (vi) above?

   Try running it for $l1 = 2, l2 = 3.5$. What do you observe?

(viii) (Optional extra) Between $\lambda = 3.9$ and $\lambda = 4$ there are several more attracting cycles. Find some of them, using the M-files to the best advantage that you can. (It is possible to be reasonably systematic in this.)

C Quadratic and exponential sequences; fixed points

Aims of the project
This is again about sequences, their limits and their attracting cycles—

sets of numbers which the terms of the sequence approach. We shall study two types of sequence, quadratic sequences (as in B above) and exponential sequences, for example x, 2^x, $2^{(2^x)}$ and so on. (Exponential sequences are treated in detail in [2], pp.406ff.) In particular we shall study 'fixed points' of sequences and relate these to their convergence behaviour. There is also a very little on Möbius sequences in this investigation; see §12.1 for the definition. You might also want to look at §12.2 which describes 'cobweb diagrams'.

Mathematical ideas used

The main mathematical idea is that of a sequence of real numbers. The theoretical part of the investigation involves calculus and manipulation of functions.

MATLAB techniques used

Existing M-files are used. You need to amend the M-file which plots cobweb diagrams of quadratic sequences so that it covers the case of 'exponential sequences'.

Quadratic sequence cobwebs

For a general discussion of cobweb diagrams, see §12.2.

The M-file `cobq.m` plots a cobweb diagram for the sequence

$$x_{n+1} = \lambda x_n(1 - x_n),$$

where λ is a fixed real number which we take to be > 0. That is, the function being used is $y = f(x) = \lambda x(1 - x)$. Try, for example, $\lambda = 3.9221934, xl = 0, xu = 1, yl = 0, yu = 1, x0 = 0.5$. (As usual, xl denotes the lower limit of x, xu the upper limit, and yl, yu are the lower and upper limits of y.) Try 20 iterations. This produces a 7-cycle of values, a so-called *superattracting* 7-cycle.

Quadratic and exponential sequences

(i) Use the M-file `cobq.m`, in the cases given by Table 12.2. Once the cobweb diagram is plotted the values of the x_i are printed out, so looking at the last ones can usually confirm your visual impression of a limit/periodic sequence/chaos (wild divergence). Table 12.2 shows the numbers to input at the prompts; n is the number of iterations.

In each case state what you consider the behaviour to be; in the case of convergence give the limit and in the case of periodic sequences, give the length of the period and the various values of x occurring in the period. Note that here a 'convergent' sequence, means one whose values

Table 12.2. *Values to try in the quadratic sequence.*

λ	x_0	xl	xu	yl	yu	n
2.5	0.5	0.5	0.7	0.5	0.7	20
3.2	0.5	0.4	0.9	0.4	0.9	20
3.4	0.5	0.4	0.9	0.4	0.9	40
3.5	0.5	0.3	0.9	0.3	0.9	20
3.8	0.5	0	1	0	1	50
3.83	0.5	0	1	0	1	20

approach a single limiting value. On the other hand a 'periodic' sequence of period k means one for which there are k numbers $l_1, \ldots, l_k$ and the values of the sequence come close in succession to these k numbers. (If you *start* with $x_0 = l_1$ say, then the sequence actually cycles round the numbers $l_2, \ldots$ and back to l_1 again. As in Investigation B, the numbers $l_1, \ldots, l_k$ are said to form an *attracting k-cycle*.) So you look at the numbers produced by the sequence and see if these 'settle down' to a single number or to a succession of the same numbers repeated.

(ii)   Amend the M-file `cobq.m` to plot the graph of $y = k^x$, where k is a fixed real number, and iterate this function, producing a cobweb diagram as before. (Call the resulting M-file `cobexp.m`.) This is just a matter of replacing the quadratic function with the function $y = k^x$ in the M-file `cobq.m`. Note that the sequence being considered here, for a fixed value of k, is x, k^x, $k^{(k^x)}$, and so on. It is called an *exponential sequence*. Remember that since in the M-file x is a vector of values, you will have to use `k.^x`

Investigate whether sequences starting at say $x_0 = 0$ are convergent or divergent, and, if convergent, what the limit is. If it depends on the value of k, then attempt to find the value(s) of k at which behaviour changes.

Here are some values to put in `cobexp.m` to get you started:

$k = 0.05$, $xl = -0.1$, $xu = 1.1$, $yl = -0.1$, $yu = 1.1$;
$k = 0.5$, $xl = -0.1$, $xu = 1.1$, $yl = -0.1$, $yu = 1.1$;
$k = 1.4$, $xl = 0$, $xu = 5$, $yl = 0$, $yu = 5$;
$k = 1.5$, $xl = 0$, $xu = 10$, $yl = 0$, $yu = 10$.

As an aid to thinking about what happens between $k = 1.4$ and $k = 1.5$, consider where the line $y = x$ meets the curve $y = k^x$. Show that this is where $k = x^{1/x}$. Sketch the graph of $y = x^{1/x}$. *Hint*: Use $x^{1/x} = e^{\ln(x)/x}$ and show that the derivative of $x^{1/x}$ is $x^{1/x}(1-\ln(x))/x^2$.

What is the largest value which this y takes? (Answer this question by using calculus to find the maximum of y.) How does this relate to the behaviour of the exponential sequence?

Fixed points

A *fixed point* of a function f is a value of x such that $f(x) = x$. Consider the functions $f(x) = (ax+b)/(cx+d)$, where $c \neq 0$, and $f(x) = \lambda x(1-x)$, where $\lambda > 0$. What are their real fixed points (if any)? Let α be a real fixed point. It is called

> *attracting* if $|f'(\alpha)| < 1$;
> *repelling* if $|f'(\alpha)| > 1$;
> *indifferent* if $|f'(\alpha)| = 1$.

The general idea is that if x_0 is close to an attracting fixed point α, then the iterates x_n will be attracted to α, that is $x_n \to \alpha$ as $n \to \infty$. Similarly if x_0 is close to a repelling fixed point α, then x_1 will be further away from α. For an indifferent fixed point various things can happen.

Except in (vii) below, you can take your starting value x_0 to satisfy $0 \leq x_0 \leq 1$.

(iii) For the case $f(x) = \lambda x(1 - x)$ show (using mathematics, not the computer!) that the only values of $\lambda > 0$ giving indifferent fixed points are $\lambda = 1$ and 3. Use the M-files to find out what happens to sequences x_n in these cases.

(iv) Show that for $f(x) = (-3x - 2)/(4x + 3)$, there are two indifferent fixed points. Use the M-files to investigate what happens to sequences in this case. More generally, show that $f(x) = (ax + b)/(cx + d)$ always has indifferent fixed points when $a + d = 0$ and $ad - bc = -1$.

(v) Investigate the fixed points and behaviour of the function given by $f(x) = (7x - 2)/(2x + 3)$.

(vi) Find one example of a Möbius sequence $(f(x) = (ax + b)/(cx + d))$ and one of a quadratic sequence $(f(x) = \lambda x(1 - x))$ with an *attractive* fixed point, and investigate with the M-files whether sequences really do have that as limit. (For the Möbius sequence, you are probably better off guessing than trying to find the condition for an attractive fixed point.)

(vii) Investigate the indifferent fixed points of $f(x) = -x$ and $f(x) = x^2 + x$. (In these cases you might want to take x_0 outside the range $0 \le x_0 \le 1$.)

13

Newton–Raphson Iteration and Fractals

Aims of the project

We shall investigate sequences of points in the complex plane which
have limits equal to roots of certain simple equations. The sequences
are generated by the complex number version of the well-known Newton–
Raphson formula. We shall in particular contrast the case of a quadratic
equation with that of a cubic equation (the contrast could not be greater!).
For a detailed discussion of the Newton–Raphson method for *real* equations, see Chapter 15. There are many semi-popular books on fractals,
for example [14].

Mathematical ideas used

Complex numbers, including solution of quadratic equations with complex coefficients are used. There is also mention of a special cubic equation. Some plane geometry is used, such as perpendicular bisectors of
segments.

MATLAB techniques used

This project uses the ability of MATLAB to draw points and lines in the
plane. The points here will represent complex numbers in the usual Argand diagram. Basic M-files are written for you, and need some amendments to apply to other examples.

13.1 Introduction

You may have met the Newton–Raphson method for approximating the
roots of an equation $f(x) = 0$. Briefly, this consists in making a guess
x_0 at a root and then refining it by $x_1, x_2, x_3, \ldots$, where

$$x_{k+1} = x_k - \frac{f(x_k)}{f'(x_k)}, k = 0, 1, 2, \ldots. \qquad (13.1)$$

164

In this project we shall apply this method to the solution of certain *complex* equations.

13.2 The equation $z^2 + 1 = 0$

As you know the above equation has solutions $\pm i$. (Since the variable is complex we shall call it z.) For this equation, the N-R formula above becomes

$$z_{k+1} = z_k - \frac{z_k^2 + 1}{2z_k} = \frac{z_k^2 - 1}{2z_k}.$$

In fact, let us write $N(z) = \frac{z^2-1}{2z}$, so $z_{k+1} = N(z_k)$. Starting with a guess z (which had better not be 0!), the successive approximations to a root are $N(z), N^2(z) = N(N(z)), N^3(z) = N(N^2(z))$, etc.

Note that we have to be careful in this project about distinguishing *iteration*, that is repeating the same function, as with N above, and *raising to a power*. If we want to write down say the fourth power of $N(z)$ then we shall write $(N(z))^4$, using brackets and the position of the 4 to remove ambiguity.

To begin with, here is some theory. You should write it out, filling in the details. We want to determine which root, if any, is found by taking the limit of numbers $N^k(z)$ as $k \to \infty$.

(i) Define $T(z)$ by $T(z) = \frac{z-i}{z+i}$ (this is of course connected with the fact that the roots of $z^2 + 1 = 0$ are i and $-i$). Verify by substituting $N(z)$ into T that $T(N(z)) = (T(z))^2$; repeated application of this shows that

$$T(N^2(z)) = (T(N(z)))^2 = (T(z))^4.$$

What will be the general result here? $T(N^k(z))$ will be *what* power of $T(z)$? (*Hint*: it is *not* the power $2k$. Try working out $T(N^3(z))$ for a start.)

(ii) Why is $|T(z)| = 1$ if and only if z is the same distance from i and from $-i$? (Recall that $|z - a|$ is the distance between the complex numbers z and a.) Let L be the set of points z with this property. What is L? Draw it on an Argand diagram of the complex numbers. It follows that points on one side of L satisfy $|T(z)| > 1$ and points on the other side satisfy $|T(z)| < 1$. Which side of L is which?

(iii) Assume that $|T(z)| < 1$. Why does it follow that $(T(z))^r \to 0$ as $r \to \infty$? (*Hint*: $|(T(z))^r| = |T(z)|^r$.) Deduce from the above that, as $k \to \infty$, we have $T(N^k(z)) \to 0$. Why does it follow that $N^k(z) \to i$?

What is the corresponding result when $|T(z)| > 1$? (*Hint:* Let $1/T(z) = U(z)$ so that $|U(z)| < 1$, then apply the same argument (check $U(N(z)) = (U(z))^2$), noting that $U^{-1}(0) = -i$.)

(iv) Now consider the two regions of the plane:

$$R_+ = \{z : N^k(z) \to i \text{ as } k \to \infty\}; \quad R_- = \{z : N^k(z) \to -i \text{ as } k \to \infty\}.$$

Draw a diagram to illustrate these regions, the line L and the roots i and $-i$. We call R_+ the *basin of attraction* for the root $+i$, and similarly R_- is the basin of attraction for the root $-i$.

Show that if z is on the set L (the common boundary of the two regions R_+ and R_-), then $N^k(z)$ stays on L for all values of k. (This is easy once you identify what L is.) So in this case iteration does not produce a root at all.

13.3 General quadratic equations

(i) The M-file `cnr1.m` (cnr is complex Newton–Raphson) takes any two complex numbers p, q and finds the quadratic equation

$$z^2 + az + b = 0 \tag{13.2}$$

with these as roots. Thus, as is well known, $a = -p - q$, $b = pq$. The Newton–Raphson formula (13.1) for this case replaces the current 'guess' z by

$$N(z) = z - \frac{z^2 + az + b}{2x + a} = \frac{z^2 - b}{2z + a}. \tag{13.3}$$

The M-file chooses a random starting point z and joins up the successive complex numbers $N(z), N^2(z), N^3(z), \ldots$ until they come very close to one of the roots p, q (marked with large crosses, one red and one green). So the zig-zag line produced shows how the method gradually approximates to a root of the equation. To choose another random starting point, press <Enter>. The M-file allows ten such starting points.

Try running the M-file and inputting $i, -i$ for the roots. You need to specify the square region of the Argand diagram shown on the screen, namely $xl \le x \le xu, yl \le y \le yu$, by giving in succession xl, xu, yl (yu is then calculated). Observe which iterations tend to i and which to $-i$; this should agree with the theory above!

Try running the M-file with, for example, $1 + 2i$, $3 + 4i$ for the roots, and a suitable region, for example, $0 \le x \le 5$, $0 \le y \le 5$. Do you have any conjecture about which starting points end up at $1 + 2i$ and which at $3 + 4i$?

(ii)   The M-file `cnr2.m` takes 500 random starting points and works out what happens to them under repeated application of the Newton–Raphson formula applied to the same quadratic equation $z^2 + az + b = 0$ with roots p, q. The roots are marked with a red cross for p and a green cross for q. Those starting points which tend to the first root (p) are plotted as small red circles and those which tend to the second root (q) are plotted as small green circles. Try the same examples as in Question (i) above, and some others. Write down the values of p and q which you use, and the limits of x and y on the screen, and sketch the line L separating the two 'basins of attraction' for each example you consider. (The basin of attraction for p is the area of the plane containing all those starting points z which give approximations tending to p.)

Can you guess what the line L separating the two 'basins of attraction' is for a general quadratic equation? Does this support your guess in Question (i)?

(iii) Guided by the following hints, find the dividing line between basins of attraction for a general quadratic equation with distinct roots p, q by a theoretical argument.

The Newton–Raphson map is still defined by equation (13.3). Let T be the transformation

$$T(z) = \frac{z - p}{z - q},$$

where as before p and q are the roots of the quadratic equation (13.2). Let L be the set of points z in the plane for which $|T(z)| = 1$, i.e. $|z - p| = |z - q|$. Draw a diagram to illustrate how L is related to p and q.

Verify that $T(N(z)) = (T(z))^2$, just as in §13.2. It follows, as there, that $T(N^k(z)) = $ *what* power of $T(z)$? The argument now splits into two cases according to the side of L on which the starting point z lies. These two sides are distinguished by $|T(z)| < 1$ and $|T(z)| > 1$. Use the same idea as in §13.2 to show that if $|T(z)| < 1$, then $T(N^k(z)) \to 0$ as $k \to \infty$ and deduce that $N^k(z) \to p$. Similarly if $|T(z)| > 1$, then (for example, by using $U = 1/T$ as in §13.2) $N^k(z) \to q$. So L is the dividing line.

Can you show that, if z lies on L, then all $N^k(z)$ also lie on L?
What happens if $p = q$? First run `cnr1.m` and `cnr2.m` for a few

choices of equal roots and observe what happens. Formulate a general result and see if you can prove it. (*Hint:* Try $T(z) = z - p$.)

(iv) Returning to the case $p \neq q$, what happens when we take our initial guess z on the dividing line L? The case $z^2 + 1 = 0$ is typical, and here L is, as you should realize by now, the x-axis ! In Question (iv) of §13.2 you showed that if z lies on L for this case, then $N(z)$ does too: the iterates $N(z), N^2(z), N^3(z), \ldots$ all lie on L.

Illustrate this by amending `cnr1.m`, calling the result `cnr3.m`, so that $p = i$, $q = -i$, the screen depicts $-10 \leq x \leq 10$, $-10 \leq y \leq 10$, and a single random starting point z on the line L is chosen. A white circle should be placed at this point and at each of the 100 subsequently calculated points by

```
plot(real(z),imag(z),'wo')
```

You should also store the values z (which will actually be real in this example as they are all on L) in a vector `xvalue`. You do this by initialising

```
xvalue=[];
```

and then after each new z is calculated using

```
xvalue=[xvalue, real(z)];
```

After the end of the calculation, plot these values against the count number by

```
plot([0:100],xvalue)
```

Print out this plot. Does is suggest to you that the values produced by the Newton–Raphson method do not converge to anything when the starting value is on L? Give a reason for your answer.

13.4 The cubic equation $z^3 - z = 0$

Now we shall look at the cubic equation $z^3 - z = 0$, which has roots $p = -1$, $q = 0$ and $r = 1$. Verify that the Newton–Raphson formula is now

$$N(z) = \frac{2z^3}{3z^2 - 1}.$$

The idea is to find the basins of attraction, that is, the regions of the plane defined by

$$R_{-1} = \{z : N^k(z) \to -1 \text{ as } k \to \infty\}$$

and similarly for R_0 and R_1 . You will see that this is rather a tall order.

Fig. 13.1. A close up view of part of the basin of attraction for the root $r = 1$ of the cubic equation $z^3 - z = 0$.

(i) Amend cnr2.m so that it applies to this example, plotting circles in three colours: a circle of a particular colour is plotted at the point (x, y) if with starting value $z = x + iy$ the N-R iteration converges to a particular root. You should take 1000 random points as the starting points. Call the resulting M-file cnr4.m.

(ii) Run the program with $xl = -2$, $xu = 2$, $yl = -2$ (so x and y go from -2 to 2). Make a rough sketch of the three basins of attraction.

Now run it for $xl = -0.6$, $xu = -0.4$, $yl = -0.1$, that is concentrating on a small portion of the earlier picture. (Mark this portion roughly on your first sketch.) Make a rough sketch of the three basins of attraction again.

The basins of attraction in the case of this very innocent-looking cubic equation are in fact infinitely complicated, in the sense that no matter how much you magnify them they still look complicated (like the Mandelbrot set). Figure 13.1 is obtained from a program just like cnr4.m except that it scans the screen pixel-by-pixel instead of taking a large number of randomly scattered points. It shows the basin of attraction for the root $r = 1$, with $xl = -0.6$, $xu = -0.4$, $yl = -0.1$. Each of the little excrescences round the main blob is infinitely complicated. The basin of attraction is a *fractal set*: however much you magnify a small portion, it never looks any simpler.

As you can see, cubics are very different from quadratics!

14

Permutations

In this chapter there are two projects. The first one (A) is about random permutations of a finite set, cycles and permutation matrices. The second (B) is an investigation of card shuffling, introducing many of the standard ideas including perfect and approximate riffle shuffles.

A Cycle decompositions

Aims of the project
We shall use MATLAB to investigate 'random' permutations, especially their disjoint cycle decompositions. There is a theoretical and experimental investigation of the [5 [5 average number of disjoint cycles occurring in a random permutation. The basic material on permutations is generally covered in a first course on abstract algebra; see, for example, [11].

Mathematical ideas used
This investigation studies permutations of a finite set, decompositions into disjoint cycles, the order of a permutation and permutation matrices. The order of a permutation involves the idea of the least common multiple (lcm) of a set of integers. Also the average number of disjoint cycles occurring in permutations of a given finite set is investigated both experimentally and theoretically. *Note*: We always write composition of permutations from right to left: the notation $\pi_2 \pi_1$ means 'do π_1 first and then do π_2'.

MATLAB techniques used
A given M-file produces 'random' permutations of the consecutive numbers $1, 2, \ldots, n$, and another breaks a permutation up into disjoint cycles. You need to amend these to calculate the lengths of cycles and the least

common multiple of the lengths. Also you need to count the total number of disjoint cycles in a permutation and to average this over a large number of trials.

A permutation of $1, 2, \ldots, n$ is a rearrangement of these numbers in some order. For example, when $n = 8$, the permutation

$$\pi = \begin{pmatrix} 1 & 2 & 3 & 4 & 5 & 6 & 7 & 8 \\ 7 & 3 & 8 & 6 & 5 & 4 & 1 & 2 \end{pmatrix} \qquad (14.1)$$

takes 1 to 7, 2 to 3, 3 to 8 and so on. The rearrangement is simply written in the bottom row, which we shall call the vector **p**. Any permutation can be written in *disjoint cycle notation*, . For π this gives $(17)(238)(46)(5)$, which means that π can be effected by leaving 5 alone; taking 4 to 6 and 6 to 4; taking 2 to 3, 3 to 8 and 8 to 2; and taking 1 to 7 and 7 to 1. The cycle (46) is called a *2-cycle* or *transposition*, (238) is a *3-cycle*, etc. Often the '1-cycle' (5) is omitted from the notation, but here we shall include it. We shall be interested in (among other things) the *total number of disjoint cycles* in the expression of a permutation. In the case of π, this number is 4. (It is reasonably clear—and true!— that the disjoint cycle decomposition of a permutation π is completely determined by π, except that the cycles may be written in a different order and each cycle may itself be cyclically permuted. For example, the above π is also equal to $(5)(382)(71)(46)$.)

We denote by $S(n)$ the set of all permutations of $1, 2, \ldots, n$. There are $n!$ of these permutations, since there are n choices for what 1 is taken to, then $n - 1$ choices for what 2 is taken to, etc., making

$$n(n - 1)(n - 2) \ldots 1 = n!$$

choices altogether.

(i) The available MATLAB function `randperm(n)` produces a 'random' permutation of the numbers $1, 2, \ldots, n$, that is, a random element of $S(n)$. Thus you type `randperm(10)` to get a random permutation of $1, 2, \ldots, 10$. Take a look at this M-file. Use your knowledge of the `sort` command to write down how `randperm.m` works. (Remember you can type `help sort` if you've forgotten about `sort` !)

Write a short M-file which creates 1000 random permutations and writes them all as the rows of a matrix bigp of size $1000 \times n$.

Hints: You will want to start with

```
n=input('Type the value of n')
```

so that you can use different values of n in different examples. Each random permutation is then generated by

```
p=randperm(n);
```

You will need a loop,

```
for j = 1:1000
```

starting *after* the input line for n. If you write, within the loop,

```
bigp(j,:)=p;
```

this makes **p** the jth row of the matrix bigp. If you do several examples in succession, it might be a good idea to type (in MATLAB)

```
>> clear bigp
```

between the examples, or add this at the beginning of your M-file.

Now take $n = 5$ and draw histograms of the *columns* of bigp. (Note that

```
hist(v,1:5)
```

draws the histogram of a vector **v** with the numbers 1,2,3,4,5 as the centres of the 'bins'.) Print out *one* of these histograms and state whether they are all reasonably level, which would suggest that the numbers 1,2,3,4,5 are equally likely to appear in any position in the permutation (and hence that `randperm.m` is a reasonably successful random permutation generator).

(ii) The M-file `cycles.m` generates a random permutation and then breaks it into disjoint cycles. It does this in exactly the same way that you would, by following numbers through the permutation until a cycle is formed and then going on to a new number to start another cycle. To help the computer realise when all the numbers from 1 to n have been 'used up' in cycles, the entries in **p** are changed to zero as they are used in a cycle. Try running the M-file a few times to see what output it produces. For one example with $n = 10$, break the permutation into cycles by hand, checking that you get the same answer as the computer does.

(iii) Amend `cycles.m` so that it counts the orders (lengths) of the cycles and then at the end finds the lcm of the lengths. It is best if the lengths are stored in a vector, say `clength`, which starts out empty:

```
clength = []
```

and is then augmented each time a new length is found:

```
clength = [clength newlength]
```

where 'newlength' stands for the length just found. To find the lcm, proceed as follows: you can assume that $l = \text{lcm}(x_1, x_2, \ldots, x_k)$ can be obtained by the sequence

$$l_1 = \text{lcm}(x_1, x_2), \quad l_2 = \text{lcm}(l_1, x_3), \quad \ldots, \quad l_{k-1} = \text{lcm}(l_{k-2}, x_k),$$

and the last number l_{k-1} calculated is the lcm l required. You can also assume that for two numbers a, b

$$\text{lcm}(a, b) = \frac{ab}{\gcd(a, b)}.$$

You can use the existing function `gcdiv.m` to find the gcd.

So the MATLAB commands needed to calculate the lcm of the lengths are

```
ord=clength(1);
for i=2:length(clength)
    ord=(ord*clength(i))/gcdiv(ord,clength(i));
end
```

and then of course you will want to display the order, here called `ord`. Give a few examples of results from this M-file.

(iv) Amend `randperm.m` so that it also calculates the corresponding *permutation matrix* A of the permutation $\pi \in S(n)$. This is defined as the matrix which starts off all zeros and then adds a 1 in the ith row at position $\pi(i)$, for $i = 1, \ldots, n$. So you first define `A = zeros(n,n)` and then have a loop

```
for i=1:n
    A(i,p(i))=1;
end;
```

(Recall that **p** is the name of the vector giving the rearrangement, the bottom row of equation (14.1) above.)

Why is it true that when you multiply A by the column vector $\mathbf{u} = (1, 2, \ldots, n)^\top$ to form $A\mathbf{u}$, you get the permutation vector **p**, written as a column? Note that in MATLAB, **u** could be written `u=[1:n]'`. It follows that powers of the permutation can be calculated by $A^2\mathbf{u}, A^3\mathbf{u}, \ldots$. Use your examples from Question (iii) above to illustrate the fact that the lcm of the lengths of the cycles equals the *order* of a permutation (that is the smallest power $k \geq 1$ for which $A^k\mathbf{u} = \mathbf{u}$). *Note:* You may find it better to produce a new version of the M-file you wrote for the earlier part of this question, in which you are able to specify the permutation instead of having it chosen randomly for you.

(v) Amend `cycles.m` to calculate also the *total number of cycles* (including 1-cycles) in the disjoint cycle expression for a permutation. Then amend it again so that it takes a given number of random permutations (which might be 1000 in practice) and calculates the *average* number of cycles which are in the disjoint cycle representations of these permutations. Run this several times with different values of n to get experimental estimates for these averages for the different n.

(vi) This is a theoretical argument which predicts the average found experimentally in Question (v) above. It involves no computing. You should fill in the details of the arguments and answer the questions.

We calculate the average E of disjoint cycles calculated over *all* $n!$ permutations in $S(n)$. Thus let $P_k(n)$ be the total number of k-cycles occurring among the disjoint cycle representations of all elements of $S(n)$. We need

$$E = \frac{P_1(n) + P_2(n) + \cdots + P_n(n)}{n!}.$$

For example, let $n = 3$. The disjoint cycle representations of the six elements of $S(3)$ are

$$(1)(2)(3), (1)(23), (2)(31), (3)(12), (123), (132).$$

Thus $P_1(3) = 6$, $P_2(3) = 3$, $P_3(3) = 2$, and $E = \frac{11}{6}$ is the average number of disjoint cycles for permutations of three objects.

How many permutations in $S(n)$ contain the particular 1-cycle (1)? The remaining $n - 1$ numbers can be permuted to anything, so there are $(n-1)!$ such permutations. Similarly for the 1-cycles $(2), \ldots, (n)$. So for each of the n possible 1-cycles there are exactly $(n - 1)!$ permutations containing that 1-cycle. So there are $n(n - 1)! = n!$ 1-cycles occurring somewhere among the complete list of permutations, expressed in disjoint cycle form, in $S(n)$. Thus $P_1(n) = n!$.

Now consider 2-cycles. Any particular 2-cycle, such as (12), occurs in exactly $(n - 2)!$ permutations in $S(n)$, and there are $\frac{n(n-1)}{2}$ possible 2-cycles (why?). So $P_2(n) = \frac{n!}{2}$.

Similarly show that $P_3(n) = \frac{n!}{3}$. (To count the number of possible 3-cycles, first count the number of ordered triples of distinct numbers, (a, b, c), all between 1 and n. Then use the fact that any particular 3-cycle comes from *three* of these, say $(a, b, c), (b, c, a), (c, a, b)$, so divide by 3.)

What is the general result for $P_k(n)$? What is the value of E? Compare this average with those calculated experimentally in Question (v) above.

(vii) Finally, consider the following slightly bizarre calculation. We are given n and a second number m, with $1 \leq m \leq n$. The disjoint cycles of any particular permutation π which contain the numbers $1, 2, \ldots, m$ might or might not actually contain all of $1, 2, \ldots, n$. For example, with $n = 4, m = 2$ consider the permutations

$$\pi_1 = \begin{pmatrix} 1 & 2 & 3 & 4 \\ 2 & 1 & 4 & 3 \end{pmatrix}, \quad \pi_2 = \begin{pmatrix} 1 & 2 & 3 & 4 \\ 1 & 3 & 4 & 2 \end{pmatrix}.$$

For $\pi_1 = (12)(34)$, the cycles containing 1,2 do *not* also contain 3,4. For $\pi_2 = (1)(234)$, they *do*.

Amend `cycles.m` to calculate the proportion of a random selection of say 1000 permutations which do have the above property, for a given m (input at the beginning, with n). *Hint*: This is a great deal easier than it sounds. You need to input m and the number of permutations to test, say `testnum` permutations. You will also need a counting variable, say `count`, to count the number of permutations which *do* have the property above. The only other essential change that needs to be made is to change

```
while i<n
```

to

```
while i<m
```

Once the cycle decomposition has been found, you need to check whether it has the required property; this is done by

```
if p == zeros(size(p))
count=count+1;
end
```

Can you see why this does the trick?

Taking say $n = 8$ and various m, conjecture a formula for this proportion in terms of n and m.

(viii) The calculation in Question (vii) can be interpreted, as follows. We have n boxes, each with its unique key which fits no other box, and each with a small slot in it through which you can push a key but can't get it back! The boxes are locked, and the keys are dropped at random into the boxes. Now boxes $1, 2, \ldots, m$ are broken open. What is the probability that every box can now be unlocked? If you can explain the connection of this with the calculation above, please do so.

B Card shuffling

Aims of the project

Card shuffling is a situation in which permutations occur in (some people's) everyday life. In this project we shall study both 'perfect' shuffles in which there is an extreme degree of regularity, and also 'rough' shuffles where there is a certain amount of chance, modelling an ordinary person's (as opposed to a card magician's) riffle shuffle of a pack of 52 cards. We shall find, surprisingly, that three (rough or perfect) riffle shuffles of a pack are by no means enough to randomise its ordering. In fact if three such shuffles are performed and then a single card is moved to another place in the pack, it is usually possible to discover the identity of that card by a simple technique of laying out the cards in columns. There is a discussion of card shuffling in [6].

Mathematical ideas used

This project involves permutations of a finite set, thought of as a pack of cards numbered $1, 2, \ldots, n$. So it is necessary to multiply permutations (that is, do one after the other). We use the 'congruence' notation: $a \equiv b \bmod m$, where a, b, m are integers and $m \neq 0$, means that $a - b = \lambda m$ for some integer λ, which can be > 0, < 0 or zero. The *order* of a permutation obtained by shuffling a pack of cards is investigated— this means the number of times a particular shuffle has to be exactly repeated so as to return the cards to their original ordering. (There is an unfortunate clash of terminology here between 'order' and 'ordering'! The latter refers to the sequence of card values top to bottom of the pack.) There is use of the disjoint cycle representation of a permutation. *Note*: We always write composition of permutations from right to left: the notation $\pi_2 \pi_1$ means 'do π_1 first and then do π_2'

MATLAB techniques used

Some amendments to existing M-files are required, adapting from odd numbers of cards to even numbers for example. The MATLAB command `sort` will come in handy.

14.1 Introduction

14.1.1 Position permutations (pps)

Consider a pile of six cards, with face values 1, 2, 3, 4, 5, 6 top to bottom. Suppose these are rearranged (shuffled) so that the face values become

4, 1, 5, 2, 6, 3 top to bottom. The corresponding *position permutation*
(pp) is

$$\pi = \begin{pmatrix} 1 & 2 & 3 & 4 & 5 & 6 \\ 2 & 4 & 6 & 1 & 3 & 5 \end{pmatrix},$$

where $\pi(x) = y$ (indicated by x in the top row and y directly underneath)
means that the card initially in *position* x from the top of the pack moves
to position y from the top of the pack.

Most shuffles are best described by their pps. For example the above
is a 'riffle shuffle' where the cards are first divided into two equal piles
which have face values 4, 5, 6 top to bottom and 1, 2, 3 top to bottom.
These two piles are then interleaved. Doing the *same* riffle shuffle again
the piles have face values 2, 6, 3 and 4, 1, 5 and the final ordering of
cards has face values 2, 4, 6, 1, 3, 5 top to bottom. You can check for
yourself that the pp corresponding to this double shuffle is simply the
square of π (do π and then do π again):

$$\pi^2 = \begin{pmatrix} 1 & 2 & 3 & 4 & 5 & 6 \\ 4 & 1 & 5 & 2 & 6 & 3 \end{pmatrix}.$$

We could now 'cut' the pack by making it into two piles 2, 4, 6 and
1, 3, 5 and reassembling so that the ordering now has face values 1, 3,
5, 2, 4, 6. The whole sequence of three shuffles (two riffles and a cut)
therefore has pp equal to

$$\begin{pmatrix} 1 & 2 & 3 & 4 & 5 & 6 \\ 1 & 4 & 2 & 5 & 3 & 6 \end{pmatrix}. \tag{14.2}$$

Note that the cut itself has pp equal to

$$\sigma = \begin{pmatrix} 1 & 2 & 3 & 4 & 5 & 6 \\ 4 & 5 & 6 & 1 & 2 & 3 \end{pmatrix}$$

(card in position 1 $\rightarrow$ position 4, etc.). You can easily verify that the
permutation in equation (14.2) is precisely $\sigma\pi^2$. Thus we can very conveniently work out the pp of the combined effect of the three shuffles by
multiplying together the pps of the individual shuffles: π then π then σ,
written $\sigma\pi^2$.

Generally:

*suppose we do a sequence of shuffles, with pps π_1, then π_2, and so on
up to π_r. Then the pp of the total shuffle is the product permutation
$\pi_r\pi_{r-1}\ldots\pi_1$. In particular, repeating a shuffle with pp π a total of r
times gives a shuffle with pp equal to π^r.*

Note that to save excessive writing we shall from now on indicate pps by their bottom line only. The top line is always $1, 2, \ldots, n$. Sometimes (§14.3) we shall be interested in the disjoint cycles representation of a pp.

Of course, ordinary playing cards have suits. We shall label them with face values $1, 2, \ldots, 52$ instead. This is the same as ordering the suits and ordering the cards A,2, $\ldots$, 10, J,Q,K within each suit.

14.1.2 Using the MATLAB 'sort' command

MATLAB has an automatic way of converting from position permutations to face values, namely the `sort` command. Suppose we start with six cards in the natural ordering 1, 2, 3, 4, 5, 6. Try the following in MATLAB.

```
>>  p=[2 6 4 1 5 3];  % Or any other list: this is the pp
>>  [v q]=sort(p);
>>  q
```

You will find that the vector **q** gives the face values of the cards left to right (which we often think of as top to bottom in a pack). Note that **v** is not used here, but it needs to be included in the `sort` command. Note that, conversely, regarding **q** as the position permutation, **p** is the vector of face values of the cards left to right: it works both ways round.

So the crucial advantage of the pp method is that by multiplying the pps of two shuffles in the ordinary way we get the pp of the combined shuffle of the cards. We can always convert to face-value orderings by using `sort`, and this is done for you in the M-files.

14.2 Ins and outs

Take a pack of n cards labelled $1, 2, \ldots, n$ and divide it into two parts, as nearly equal as possible. A *perfect riffle shuffle* is obtained by alternating cards from the two half-packs. There are four cases, according to whether n is even or odd, and according to which card goes on top.

14.2.1 Even in-shuffles

If n is even, say $n = 2k$, then the two half-packs contain $1, 2, \ldots, k$ and $k+1, k+2, \ldots, 2k$. If card $k+1$ goes on top this is called an *in-shuffle*

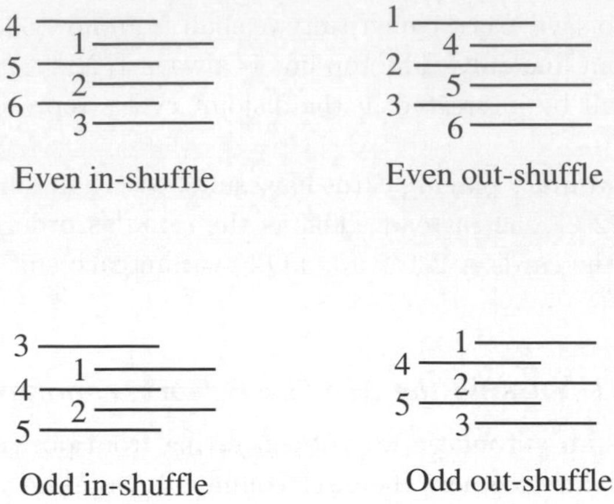

Even in-shuffle   Even out-shuffle

Odd in-shuffle   Odd out-shuffle

Fig. 14.1. In- and out-shuffles for an even and odd number of cards, which start out top to bottom in the order $1, 2, 3, \ldots$..

and the face-values of the cards are now

$$k + 1, 1, k + 2, 2, k + 3, 3, \ldots, 2k, k.$$

See Figure 14.1 for the case $k = 3$.

Write down the corresponding position permutation, π say, and verify (e.g. by a couple of examples like $n = 6$, $n = 10$) that π can be written

$$\pi(x) \equiv 2x \pmod{2k + 1},$$

where recall that this means $\pi(x) - 2x$ is an exact multiple of $2k + 1$.†
In this case $\pi(x)$ will either equal $2x$ or $2x - (2k + 1)$.

Note that since this is the *position permutation* we know immediately that doing the same shuffle again results in a pp obtained by doing π twice:

$$\pi^2(x) = \pi(\pi(x)) \equiv \pi(2x) \equiv 4x \pmod{2k + 1}. \qquad (14.3)$$

Thus here the card which starts in position x before any shuffling is done ends up, after two shuffles, in position $4x \pmod{2k+1}$. Verify this directly in the case $n = 6$.

What is the smallest number of times you need to repeat this shuffle, for $n = 6$, to get the cards back in their original ordering?

† The only properties of $\equiv$ mod n we need are: if $a \equiv b$ and $c \equiv d$, then (i) $a \pm c \equiv b \pm d$ and (ii) $ac \equiv bd$. In particular if $a \equiv b$, then $a^r \equiv b^r$ for any integer $r > 0$.

To illustrate how to combine permutations in an M-file, the M-file `riffle1.m` repeats the in-shuffle, for n even, any number of times. Study this M-file: you will need to amend it later. Note in particular that it uses `sort` to print out the final face-value ordering of the cards, since that is what we want to know. Use it to find the smallest number of in-shuffles of 14 cards which are needed to get them back into the original ordering.

Now look at the M-file `riffle1a.m` . This keeps multiplying 2 by itself and reducing the result mod $2k + 1$, that is, taking the remainder on dividing the result by $2k + 1$. Try $k = 7$; the value of r which it produces should be the same as the number of in-shuffles of 14 cards above.

The explanation for this is as follows. By repeating the argument leading to equation (14.3) we find that the position permutation for r repetitions of the shuffle is the permutation π^r which has the property that for each position x,

$$\pi^r(x) \equiv 2^r x \pmod{2k + 1}.$$

It follows that if $2^r \equiv 1$, then every card is back in its original position! Conversely if every card is back in its original position then in particular the card in position 1 is back to position 1, so $2^r \equiv 1$. So the smallest number of repetitions of the shuffle which return all cards to their original positions is the smallest r with $2^r \equiv 1 \pmod{2k + 1}$. This smallest number is called the *order* of the shuffle, or the *order* of the permutation giving the shuffle. (The terminology is of course very unfortunate! Order*ings* of cards say what the face values are top to bottom of the pack.)

Use `riffle1a.m` to find this smallest r for all even numbers from 40 to 52 inclusive.

Your task in what follows is to imitate the above analysis in three other cases. Because these are slightly different, some hints will be given.

14.2.2 Even out-shuffles

Here there are $n = 2k$ cards but 1 goes on top, so the face-value ordering of the cards after one shuffle is

$$1, k + 1, 2, k + 2, 3, k + 3, \ldots, k, 2k$$

See Figure 14.1 for the case $k = 3$. Verify by looking at say $n = 6, n = 10$ that the position permutation in this case is π, where

$$\pi(x) \equiv 2x - 1 \ (\text{mod } 2k - 1) \text{ for } 1 \leq x \leq 2k - 1,$$

while $\pi(2k) = 2k$. There is one additional complication: for $x = k$, the position is $2k - 1$, not 0. So in working out $\pi(x)$ we want to take the remainder of $2x - 1$ on division by $2k - 1$, but, since there is no position 0, remainder 0 means position $2k - 1$.

To make this easier, a special function called `remm.m` has been created for you. This equals the ordinary `rem` unless the remainder is 0. Thus

   `remm(19,11)` gives 8,

but

   `remm(22,11)` and `remm(11,11)`

give 11.

First, use the function `remm.m` to modify `riffle1.m` to handle the case of even out-shuffles. You will need to change the i-loop to go only to $2k - 1$ and add `store(2*k)=2*k;` after the end of the i-loop but before `position_perm=store`. Call your modified M-file `riffle2.m` and use it to verify that for 52 cards you return to the original ordering after only *eight* out-shuffles. How does this compare with the number of in-shuffles for 52 cards which you found using `riffle1a.m` above?

Next, fill in the details of the following sketch to show that *the number of in-shuffles required to return $2k$ cards to their original positions is the smallest r for which*

$$2^r \equiv 1 \ (\text{mod } 2k - 1).$$

Here is the sketch of an argument. Note that the card starting in position $2k$ stays there after any number of out-shuffles, so we exclude $x = 2k$ in what follows. As before, combining shuffles just amounts to multiplying position permutations in the usual way, so two shuffles produces a position permutation π^2, where

$$\pi^2(x) \equiv 2(2x - 1) - 1 = 4x - 3 \ (\text{mod } 2k - 1).$$

By induction it follows that for r shuffles we get pp equal to π^r, where

$$\pi^r(x) \equiv 2^r x - (2^r - 1) \ (\text{mod } 2k - 1).$$

Hence, *if* $2^r \equiv 1$, *then* π^r returns every card to its original position, since $\pi^r(x) \equiv x$ for all x. Conversely, *if* every card is returned to its original position, *then* the card in position 2 is returned, so that $2^{r+1} - 2^r + 1 \equiv 2$, i.e. $2^r \equiv 1$. So the smallest number of in-shuffles returning every card to its original position is the smallest r such that $2^r \equiv 1 \pmod{2k-1}$.

Use this and a suitable amendment to `riffle1a.m` (call it `riffle2a.m`) to find the smallest number of in-shuffles returning every card to its original position, for every even number $n = 2k$ from 40 to 52 inclusive.

14.2.3 Odd packs of cards

If the number of cards is odd, say $2k+1$, then we split the pack into two, having k and $k+1$ cards in the two parts. The two parts are riffled together in such a way that the top and bottom cards of the $k+1$ part become the top and bottom cards of the combined pack (the $k+1$ cards 'straddle' the k cards). There are two choices, according as the part with k cards consists of those with face values $1, \ldots, k$ (in-shuffle) or $k+2, \ldots, 2k+1$ (out-shuffle). See Figure 14.1 for the case $k = 2$. Verify that the corresponding position permutations are given by respectively

$$\pi(x) \equiv 2x \pmod{2k+1} \quad \text{and} \quad \pi(x) \equiv 2x - 1 \pmod{2k+1}. \quad (14.4)$$

Adapt `riffle1.m` to these two cases (calling the M-files `riffle3.m` and `riffle4.m`) and find the smallest number of in- and out-shuffles which return 15 cards to their original ordering. Note that you will want to use `remm.m` rather than `rem` to avoid having a zero remainder.

Show that for both in- and out-shuffles, the smallest number of shuffles to return the cards to their original ordering is the smallest value of r such that $2^r \equiv 1 \pmod{2k+1}$. Use `riffle1a.m` to find the smallest number r for odd packs of cards with $n = 2k+1$ from 39 to 51 inclusive.

14.2.4 Riffles and cuts for odd packs

There is a very interesting phenomenon which holds for odd packs and which we shall investigate theoretically here. No computing is required, but of course you are at liberty to write some illustrative M-files if you wish!

A *cut* of a pack is a *cyclic permutation*, that is, it preserves the 'cyclic' ordering of the cards, which is what you get by writing the face-values round a circle. For example if the pack 1, 2, 3, 4, 5 is cut below 3 the new face-value ordering is 4, 5, 1, 2, 3. If these numbers are written round a circle then the ordering is the same as it was at the beginning. In general, for n cards, explain by giving some examples why cutting below the card in position c below the top produces a position permutation

$$n - c + 1, n - c + 2, \ldots, n - 1, n, 1, 2, \ldots, n - c - 1, n - c.$$

Let us call this permutation σ_c, so that $\sigma_c(x) \equiv x - c \pmod{n}$.

Now assume n is odd and equal to $2k + 1$. According to equation (14.4), a perfect riffle of n cards is given by the position permutation π where $\pi(x) \equiv 2x$ or $2x - 1 \pmod{n}$. Show that, in both cases,

$$\pi\sigma_c = \sigma_{2c}\pi.$$

Here we are just multiplying permutations in the usual way, so the equation states that

$$\pi(\sigma_c(x)) \equiv \sigma_{2c}(\pi(x)) \pmod{n} \text{ for all } x.$$

Show that both sides are congruent to $2x - 2c \bmod n$. Explain why this implies that when riffles are interspersed with cuts, the same effect can be achieved by doing a sequence of riffles *first* and *then* a sequence of cuts. Furthermore, it is pretty clear that doing any sequence of cuts one after the other is the same as doing a single cut. Thus a sequence of riffles and cuts has pp of the form

$$\sigma\pi_r\pi_{r-1}\ldots\pi_1,$$

where σ is a cut and the π_i are (in- or out-)shuffles.

Now think back to the smallest number of times, r that (say) an in-shuffle of the $n = 2k + 1$ cards must be repeated to return the cards to their original ordering (§14.2.3). Suppose that r in-shuffles (for this r) are interspersed with any number of cuts. It follows from the above that the overall position permutation can be written as $\sigma\pi^r$, where σ is some cut. Why does it follow that the cards are now in the same cyclic ordering as they started?

Show that the same result holds for out-shuffles.

It is slightly harder to deduce that the same holds for any mixture of in- and out-shuffles, so long as there are r of them. (A sketch is given in §14.5 below.) Thus *riffle shuffles and cuts are a very ineffective way of randomising an odd pack of cards:* they always return to their original cyclic ordering after a relatively small number of shuffles. The situation for even packs, incidentally, is completely different. It can be shown that a suitable sequence of riffles and cuts can produce *any* of the $n!$ permutations of $n = 2k$ cards, so riffles and cuts are an effective way of randomising even packs.

14.3 Cycles

Like any other permutation, the position permutation corresponding to a riffle shuffle can be broken into disjoint cycles. An M-file which does this for even in-shuffles has been written for you: it is called `riffle1c.m`. It displays the cycles and the cycle lengths. For even numbers from 40 to 52 use `riffle1c.m` and the known fact that the order of a permutation is the lcm of the lengths of its disjoint cycles to verify your results in §14.2.1. For each even number, write down the lengths of the cycles and the lcm. Note that 52 cards have a special property: the even in-shuffle is a single cycle. What is the next even number with this property?

14.4 Rough riffles (ruffles)

14.4.1 One ruffle

Unless you are extremely skilled, a perfect riffle shuffle is beyond your powers. So here we consider how we might model an ordinary person's attempt at a riffle, which we shall call a rough riffle, or *ruffle* for short. See Figure 14.2. To fix ideas we shall consider an even number of cards, $n = 2k$, which is initially divided into two parts, with m and $2k - m$ cards, where we shall assume that $k \leq m \leq k + 3$: an almost perfect cut of the cards.

The m cards at the top of the pack (when we start with a new pack these are labelled $1, 2, \ldots, m$) are now ruffled into the the remaining cards. There are $2k - m + 1$ 'slots' into which a card from the top m cards can go, namely the slot above card $m + 1$ and the slots below cards $m + 1, m + 2, \ldots, 2k$. We shall assume that the ruffle puts 0, 1, 2 or 3 cards into each one of these slots.

We try now to make a reasonable guess for the probabilities of 0, 1, 2 or 3 cards being put into a particular slot. Let p_i be the probability of i cards going into a slot, for $i = 1, 2, 3, 4$. Then we certainly want

$$p_0 + p_1 + p_2 + p_3 = 1.$$

The 'expected' number of cards to go into a slot is $p_1 + 2p_2 + 3p_3$ and there are $2k + m + 1$ slots, so the 'expected' number of cards used up is $(2k - m + 1)(p_1 + 2p_2 + 3p_3)$. Perhaps we should equate this with the number of cards available to go into the slots, which is m. Thus we shall try

$$(2k - m + 1)(p_1 + 2p_2 + 3p_3) = m.$$

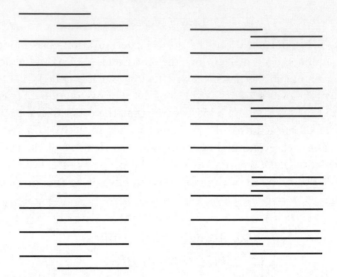

Fig. 14.2. A perfect riffle shuffle of 22 cards (left) and a rough riffle shuffle (ruffle) of the cards (right).

Thus if p_2 and p_3 are given (as well as k and m) then p_0, p_1 are given by

$$p_1 = \frac{m}{2k - m + 1} - 2p_2 - 3p_3, \quad p_0 = 1 - p_1 - p_2 - p_3. \qquad (14.5)$$

Of course, we need to choose p_2, p_3 so that p_0, p_1 are positive and < 1. Replacing the fraction in equations (14.5) by 1 (which is a reasonable approximation when k is large) show that we need only choose p_2, p_3 with $2p_2 + 3p_3 \leq 1$. For large k this will be a guide to choosing p_2, p_3.

What do you think are reasonable values for p_2, p_3? For $n = 2k = 52$, an ordinary pack, what does this give for p_0, p_1, taking the four possible values of m in turn?

An M-file has been written for you, called `ruffle1.m`, which implements the above model of a ruffle. It is rather complicated since it has to take care of various possibilities such as the cards $1, 2, \ldots, m$ running out before the slots, or vice versa. If a negative probability p_0 or p_1 is detected, then the M-file simple ends.

Run the M-file `ruffle1.m` several times, using what you consider reasonable probabilities, and taking $k = 26$ (an ordinary pack). Perhaps you can take one of these and show how the cards from one (rough) half have been slotted into the gaps between cards of the other half, using a suitable diagram.

The M-file `ruffle1c.m` finds the cycles for a ruffle, as in §14.3 above. Run this three times and for each one calculate the order of the ruffle, which is the lcm of the lengths of the cycles. This equals the number of times this same ruffle would have to be repeated to get the cards back to their original ordering. The only point here is that you might find the order very different from that of a perfect riffle of 52 cards.

14.4.2 Two or more ruffles

The M-file `ruffle2.m` is more elaborate in that it does a sequence of ruffles. In your experiments you will use up to four ruffles. The idea is to discover whether 2, 3 or 4 ruffles have the effect of 'randomising' the pack. Having performed the ruffles, you are invited to select a single card, the ith from the left after the ruffle (in the pack this would be the ith card from the top), and move it to the jth position from the left, where i and j should not be too close together. (*Suggestion:* to simplify your experiments, stick to $i = 1$: the first card is moved somewhere else.) Note down the face value of the card which you have moved. We shall see whether it is possible to detect which card was moved; if so, this is a sure sign that the pack has *not* been randomised.

After moving one card, the cards are laid out in columns in a certain way. By running the M-file with some fairly small values for k (you *might* have to be very careful about your values for p_2, p_3 if k is small), see if you can discover the method used for laying out the cards in columns. Ignore all -1s in the columns, they are merely place fillers.

With luck, the moved card will appear in a column by itself (apart from -1's below it). Experiment with two, three or four ruffles and report your results on whether the moved card can be detected. Try it out with a real pack!

Here is the output from a typical run, using $p_2 = 0.2$, $p_3 = 0.05$ and three ruffles. The card numbered 28 was moved to the tenth position (so [i, j] is [1 10]), and, as you can see, it stands out like a sore thumb!

ans =

1	22	39	29	35	45	10	28	17
2	23	40	30	36	46	11	-1	18
3	24	41	31	37	47	12	-1	19
4	25	42	32	38	48	13	-1	20
5	26	43	33	-1	49	14	-1	21

6	27	44	34	-1	50	15	-1	-1
7	-1	-1	-1	-1	51	16	-1	-1
8	-1	-1	-1	-1	52	-1	-1	-1
9	-1	-1	-1	-1	-1	-1	-1	-1

14.5 Appendix

This is a sketch proof of a result in §14.2.4. All permutations π_i here are in- or out-shuffles for n odd. We perform any sequence of in- or out-shuffles, interspersed with cuts. As in §14.2.4, this is equivalent to doing all the shuffles first and then a cut σ.

The idea is to prove that if we choose r with $2^r \equiv 1 \bmod n$, then any sequence of r in- or out-shuffles, interspersed with cuts, produces merely a cut, that is, a cyclic permutation of the n cards. We also prove conversely that for this to happen after r in- or out-shuffles interspersed with cuts, we must have $2^r \equiv 1 \bmod n$.

Now $\pi_1(x) = 2x$ or $2x - 1$, $\pi_2\pi_1(x) = 4x$ or $2(2x - 1)$ or $2(2x)$ or $2(2x - 1) - 1$, that is, $4x$ or $4x - 1$ or $4x - 2$ or $4x - 3$. In general we have

$$\pi = \pi_r\pi_{r-1}\ldots\pi_1(x) \equiv 2^r x + l,$$

where l can be any of $0, 1, 2, \ldots, 2^r - 1$. Here and below, $\equiv$ means $\equiv$ mod n.

Suppose $2^r \equiv 1$. Then $\pi(x) \equiv x + l$ for all x and a fixed l, and this is a cyclic permutation, that is, a cut. Hence $\sigma\pi$ is also a cut.

Conversely suppose that $\sigma\pi$ is cyclic, so that π is cyclic. Then

$$2^r x + l \equiv x + l'$$

for all x and some fixed l, l'. Write $l' - l = m$, so that $2^r x \equiv x + m$. For $x = 1$ this gives $2^r \equiv 1 + m$ and for $r = 2$ it gives $2^{r+1} \equiv 2 + m$. Subtracting these gives $2^r \equiv 1$.

15

Iterations for Nonlinear Equations

One can find a lot of mathematical examples involving iterations and the solution of nonlinear equations. Such equations arise from solving nonlinear problems — either differential or statistical. Real world problems are often nonlinear and may involve more than one independent variable, although techniques may resemble or reduce to the one dimensional case. The main reason for iterations is that direct (i.e. analytical) solutions of nonlinear equations are in general difficult to find, and numerical solutions need more than one step to converge. This chapter mainly considers equations with real coefficients; see Chapter 13 for the complex case.

Aims of the project
The purpose of this investigation is to study if, when and how numerical methods work in the context of solving nonlinear equations. In particular, the important issues of accuracy and convergence speed of iterative methods are considered.

Mathematical ideas used
This project involves vectors and matrices. The method is based on iterations and linearisations of nonlinear equations. We first consider one equation in one unknown (prefixed by 1D for simplicity), and then consider systems of equations in multiple unknowns (prefixed by 2D). Two M-files `full_new.m` and `gauss_ja.m` together with `f_rate.m` are used throughout this chapter.

MATLAB techniques used
This project is about solving equations. Two M-files `full_new.m` and `gauss_ja.m` will be used for such a purpose. To use them, you need to supply the equation information by a simple M-file. To analyse the

convergence speed, use the M-file `f_rate.m` and a visualisation M-file `cont4.m` is given as an example.

Throughout this chapter, superscripts are used for sequences while subscripts are reserved for coordinates in the case of multiple variables.

15.1 1D: Method 1 — Newton–Raphson

Write a single equation as

$$F(x) = 0.$$

The Newton–Raphson method takes the form:

$$x^{(n+1)} = x^{(n)} - F(x^{(n)})/F'(x^{(n)}),$$

where F' is the first derivative of function F. Given an initial guess $x = x^{(0)}$ for the root, we hope that the sequence of iterates is getting closer and closer to the true solution, that is, $x^{(n)} \to x = x^*$ as $n \to \infty$.

For your convenience, we have developed an M-file `full_new.m` implementing the Newton method. To illustrate, consider the solution of this specific example:

Example 1

$$F(x) = x^3 - 10x^2 + 27x - 18 = 0, \tag{15.1}$$

given that the root is approximately near $x^{(0)} = 7.4$ (note: the exact solution is $x^* = 6$). As the first derivative $F'(x) = 3x^2 - 20x + 27$, we can prepare a file (say) `f_ex1n.m` which contains the following†

```
function [F,J] = fun_name ( P ) ; %  Example 1
x = P(1) ;                        %  (Newton-Raphson)
F = [x^3-10*x^2+27*x-18];         %  The equation
J = [3*x^2 - 20*x + 27 ];         %  The Jacobian
```

Then the MATLAB command is simply

```
>>  Root = full_new('f_ex1n', 7.4)              % Form A
>>  Root = full_new('f_ex1n', 7.4, tni)         % Form B
>>  Root = full_new('f_ex1n', 7.4, tni, tol)    % Form C
>>  Root = full_new('f_ex1n', 7.4, tni, tol, hi)% Form D
```

† For users of MATLAB V3.5, it should be noted that the second line $x = P(1)$ in the M-file `f_ex1n.m` must be modified to $x = P$, and the same modification is needed in `f_ex1g.m` later.

where *tni* is the total number of iterations requested (say 12) starting from $x^{(0)} = 7.4$, *tol* is the stopping tolerance for $|F(x)| < tol$ (say 1.0E$-$4) and *hi* specifies if intermediate iterates are shown ($hi = 1$ for 'yes', $hi = 0$ for 'no'). With Forms A–C, only the final iterate is displayed (i.e. $hi = 0$) and variables *tni* and *tol*, if not specified, take the default values of 20 and 1.0E$-$3 respectively.

15.2 1D: Method 2 — Gauss–Jacobi

The Gauss–Jacobi method rearranges a given equation $F(x) = 0$ as $x = G(x)$ and then forms the iteration

$$x^{(n+1)} = G(x^{(n)}), \qquad \text{for } n = 0, 1, 2, \ldots.$$

This method has been coded into the M-file `gauss_ja.m`. Its usage is similar to `full_new.m` differing only in preparation of a function M-file.

Thus Example 1 above can be solved by (similar to `full_new.m`)

```
>>  Root = gauss_ja('f_ex1g', 7.4)                % Form A
>>  Root = gauss_ja('f_ex1g', 7.4, tni, tol, hi) % Form D
```

where a shorter function M-file `f_ex1g.m` contains

```
    function G = fun_name( P ); % Example 1
    x = P(1);                   %              (Gauss-Jacobi)
    R = 18 - x^3 + 10*x^2;
    G = R / 27;                 % The following is also ok
    % R = 10*x^2-27*x+18;       % Intermediate quantity
    % G = sign(R)*abs(R)^(1/3); % use SIGN to avoid complex
```

where line one contains no [ ]. In fact, you should find that the above choice of G gives rise to poor convergence.

Note $G(x)$ is not unique. We can even view the Newton method as a Gauss–Jacobi method with the particular choice $G(x) = x - F(x)/F'(x)$. For the above Example 1, $F(x) = x^3 - 10x^2 + 27x - 18 = 0$ has been rewritten as $x = G(x)$ with $G(x) = (18 + 10x^2 - x^3)/27$, but this $G(x)$ is not unique as $G(x) = \left(10x^2 - 27x + 18\right)^{1/3}$ is another possibility.

15.3 1D: Convergence analysis

Neither method 1 nor 2 works well all the time. When a method does not work, there are two possible reasons: (1) the method is no good and in this case we have to try something else; (2) the initial guess is too far away from a true root and this problem can be cured to some extent. When a method does not work well, it usually means slow convergence — a situation similar to (1) above — and is a sign of an inferior method. Below we hope to demonstrate these problems with some analysis.

15.3.1 Search region

By 'search region', we mean the region that contains the required root(s). $\mathbf{R} = (-\infty, +\infty)$ can be a rearch region but we hope to do better than being content with this large region. As most equations come from practical problems, we may choose a 'search region' based on *a priori* knowledge; for example $[1, 10]$ can be a good search region when x represents the height of people in feet.

In the present one-variable case, i.e. $F(x) = 0$, we may get some idea of the roots of $F(x)$ in an interval $[a, b]$ by using the MATLAB command `plot`. Consider

Example 2

$$F(x) = (2 - \sin x)(x^3 - 10x^2 + 27x - 18) = 0. \qquad (15.2)$$

For this example, try

```
>>   x = 0.5 : 0.01 : 6.5;
>>   F =  (2-sin(x)) .* (x.^3 - 10*x.^2 + 27*x - 18) ;
>>   plot(x,F, x,zeros(size(x)) ) ;    grid
```

to get a graph like Figure 15.1. Then it can be observed that all roots are in (say) $[0, 7]$ and so $[0, 7]$ is a good search region for the initial guess $x = x^{(0)}$.

15.3.2 Convergence region

The convergence region for a method is defined to be the collection of all initial points of a search region from which a convergent solution can be obtained. Obviously a convergence region is embedded in the search region. To find such a convergence region, we have to start from 'all' points of a search region and record those successful initial points. In practice, consider only those points in some step-length.

For instance, with the search region $[0, 8]$ for Example 1, we may start in turn at $x^{(0)} = 0$, $1/8$, $2/4$, $\ldots$, $8 - 1/8$, 8; here a step-length of $\Delta x = 1/8 = 0.125$ is used. In MATLAB notation, this means that

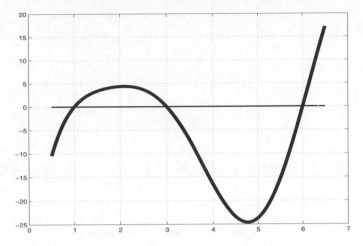

Fig. 15.1. Decide on a search region.

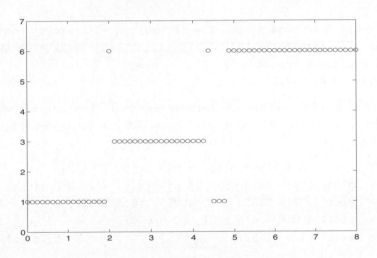

Fig. 15.2. Find a convergence region of three intervals.

x0 = 0 : 0.125 : 8. Furthermore, we can prepare an M-file as follows to generate a graph like Figure 15.2:

```
tni = 30 ; X = [] ;   R = [];
for x0 = 0 : 1/8 : 8 %%%% f_ex1n.m for Example 1 %%%%
    root = full_new('f_ex1n', x0, tni) ;
```

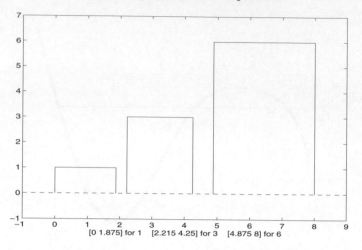

Fig. 15.3. Example 2: convergence region for roots $x = 1, 3, 6$.

```
    if 0 <= root <= 8,   X = [X; x0]; R = [R; root]; end
end %%%%%%%%%%%%%%%%%%%%%%%%%%%%%%%%%%%%%%%%%%%%%%%%%%%%%%%
Convergence_region = [X  R],      plot(X, R, 'o')
axis([0 8  0 7])
```

Apart from a few 'odd' points, there are essentially three intervals which make up a convergence region. So Figure 15.2 can be presented better in the form of Figure 15.3 which is produced by

```
>>   x = [0 0 1.875 1.875]; y =[0 1 1 0];
>>   plot(x,y);  hold on   %% plot(x,y,'o'); % Optional
>>   plot([2.215 2.215 4.25 4.25], [0 3 3 0]);
>>   plot([4.875 4.875 8 8], [0 6 6 0]);
>>   axis([-1 9 -1 8]);                              hold off
```

15.3.3 Convergence order

Different iterative schemes converge to the solution at different rates or speeds, assuming that they converge. These rates provide a measure of the performance of each method. The rate of convergence of a method is characterised by its *convergence order*. If a sequence $x^{(0)}, x^{(1)}, \ldots,$ $x^{(n)}, \ldots$ satisfies $|x^{(n+1)} - x| \propto |x^{(n)} - x|^k$, that is,

$$|x^{(n+1)} - x| = C|x^{(n)} - x|^k, \qquad n = 0, 1, 2, \ldots$$

that is,

$$\begin{aligned}
|x^{(1)} - x| &= C|x^{(0)} - x|^k \\
|x^{(2)} - x| &= C|x^{(1)} - x|^k \\
&\cdots \\
|x^{(n+1)} - x| &= C|x^{(n)} - x|^k \\
&\cdots
\end{aligned}$$

for some constants C and k (independent of n), then the method that produces this sequence of iterates is of *convergence order k*. Here x is a solution of $F(x) = 0$, $\propto$ means 'proportional' and $|\cdot|$ denotes the absolute value (modulus).

The higher the value of k, the faster is the convergence and the better is the numerical method being used. To estimate constant C and order k, we need at least three iterates and the exact root, or four iterates with the last one used as the 'exact' root.

The M-file: f_rate.m

If the complete sequence is more than four iterates, we may use the M-file f_rate.m to calculate estimates of its convergence order k. As every three iterates plus the very last one produce an estimate, the M-file f_rate.m offers a choice of printing all estimates or just an average. The syntax is simply:

```
>>  k = f_rate (root, hi)
```

where root is the sequence of iterates that are obtained from full_new.m or gauss_ga.m, and $hi = 1$ asks for all estimates to be recorded (while $hi = 0$ only requires the averaged and final k estimate to be recorded).

For Example 1, we can call f_rate.m as follows:

```
>>  root = full_new('f_ex1n', 7.4, 20, 0.0001, 1) ;
>>  k = f_rate(root, 1)
>>  mean(k)
```

where mean finds an average value. As a final example, one may explicitly specify root

```
>>  root = [ 3.46   3.5   3.61   3.735   3.755]'
>>  K = f_rate(root, 1)
    K = 3.8743   3.5091      %% Order estimates
>>  A = mean(K)
    A = 3.6917                %% Average
```

15.4 2D: Iterations for nonlinear systems

In the remaining part of this chapter, we study iterative methods for systems of equations. The two M-files `full_new.m` and `gauss_ja.m` together with `cont4.m` and `f_rate.m` will be used.

Denote a system of m nonlinear equations (in vector notation) as $\mathbf{F}(\mathbf{x}) = \mathbf{0}$, where

$$\mathbf{F} = (F_1, F_2, \ldots, F_m)^\top \qquad \text{and} \qquad \mathbf{x} = (x_1, x_2, \ldots, x_m)^\top.$$

Denote the Jacobian matrix of $\mathbf{F}$ by J, that is, the matrix of all first derivatives with $J_{ij} = \frac{\partial F_i}{\partial x_j}$. The essential tool we use is the Taylor theorem that is usually covered in a calculus course.

Given a vector function $\mathbf{F}(\mathbf{x})$ with $m = 2$, i.e. $\mathbf{x} \in \mathbf{R}^2$, each component of $\mathbf{F}(\mathbf{x})$ can be expanded in the Taylor series, that is,

$$\begin{cases} F_1(x_1, x_2) = F_1(x_1^{(0)}, x_2^{(0)}) + (x_1 - x_1^{(0)})\dfrac{\partial F_1}{\partial x_1} + (x_2 - x_2^{(0)})\dfrac{\partial F_1}{\partial x_2} + hot, \\[2mm] F_2(x_1, x_2) = F_2(x_1^{(0)}, x_2^{(0)}) + (x_1 - x_1^{(0)})\dfrac{\partial F_2}{\partial x_1} + (x_2 - x_2^{(0)})\dfrac{\partial F_2}{\partial x_2} + hot, \end{cases}$$

that is,

$$\mathbf{F}(\mathbf{x}^{(1)}) = \mathbf{F}(\mathbf{x}^{(0)}) + J \cdot \left[\mathbf{x}^{(1)} - \mathbf{x}^{(0)} \right] + hot,$$

where *hot* stands for 'higher order terms' and J is the Jacobian matrix evaluated at $\mathbf{x} = \mathbf{x}^{(0)}$

$$J = J(\mathbf{x}^{(0)}) = \begin{pmatrix} \dfrac{\partial F_1}{\partial x_1} & \dfrac{\partial F_1}{\partial x_2} \\[3mm] \dfrac{\partial F_2}{\partial x_1} & \dfrac{\partial F_2}{\partial x_2} \end{pmatrix}.$$

Further the famous Newton–Raphson method can be derived.

15.4.1 2D: Method 1 — Newton–Raphson

In the general case of m dimensions, i.e. $\mathbf{x} \in \mathbf{R}^m$, the Taylor theorem gives a linear approximation to $\mathbf{F}(\mathbf{x})$ in the vicinity of any point $\mathbf{x}^{(0)}$. If this point $\mathbf{x}^{(0)}$ is an initial guess close to the true solution $\mathbf{x} = \mathbf{x}^*$, *hoping that the linear approximation is close to* $\mathbf{F}(\mathbf{x})$, we may ignore the higher order terms and solve for the 'root' of this linear equation giving

$$\mathbf{x}^{(1)} = \mathbf{x}^{(0)} - J^{-1}(\mathbf{x}^{(0)})\mathbf{F}(\mathbf{x}^{(0)}),$$

where the *Jacobian matrix* J is evaluated at $\mathbf{x} = \mathbf{x}^{(0)}$. Repeating the above process, we obtain the Newton–Raphson method

$$\mathbf{x}^{(n+1)} = \mathbf{x}^{(n)} - J^{-1}(\mathbf{x}^{(n)})\mathbf{F}(\mathbf{x}^{(n)}), \qquad\qquad n = 0, 1, 2, \ldots$$

where the *Jacobian matrix* J is evaluated at $\mathbf{x} = \mathbf{x}^{(n)}$.

As with 1D, we hope the sequence of iterates is getting closer and closer to the true solution as $n \to \infty$. Here don't be put off by the matrix J, as it only plays the role of $F'(x^{(n)})$ in the single variable case; so $J^{-1}(\mathbf{x}^{(n)})$ corresponds to $1/F(x^{(n)})$.

Now consider the solution of the following two equations, that is, $\mathbf{F}(\mathbf{x}) = (F_1, \ F_2)^\top = \mathbf{0}$ with the unknown $\mathbf{x} = (x, \ y)^\top$

Example 3

$$\begin{cases} F_1(x,y) = \sin x - y^3 - 8 = 0, \\ F_2(x,y) = x + y - 3 = 0. \end{cases} \tag{15.3}$$

The Jacobian matrix is

$$J = (J_{ij}) = \begin{pmatrix} \frac{\partial F_1}{\partial x} & \frac{\partial F_1}{\partial y} \\ \frac{\partial F_2}{\partial x} & \frac{\partial F_2}{\partial y} \end{pmatrix} = \begin{pmatrix} \cos x & -3*y^2 \\ 1 & 1 \end{pmatrix}.$$

Again given an initial guess for the root(s), say $\mathbf{x}^{(0)} = (\ 4.7, \ -1.9 \ )^\top$, we can use the Newton–Raphson iterations to find the exact solution $\mathbf{x} = (5.075, \ -2.075)^\top$.

The iteration formula can be written as $\mathbf{x}^{(n+1)} = \mathbf{x}^{(n)} - \mathbf{d}^{(n)}$, where $\mathbf{d}^{(n)}$ is the solution of the linear equations $J(\mathbf{x}^{(n)})\mathbf{d}^{(n)} = \mathbf{F}(\mathbf{x}^{(n)})$. The rewritten iteration formula is more efficient as it avoids using, or calculating, the inverse of matrix J (see Chapter 16). Of course for the single variable case, there is no difference; refer to §15.1. As we know, the MATLAB command for finding $\mathbf{d}^{(n)}$ is simply $d = J \setminus F$. This is how we have implemented the Newton method in the following M-file.

The M-file: full_new.m

To solve Example 3, prepare the following function M-file f_ex2n.m

```
function [ F, J ] = anyname ( P ); % Ex 3 - multiple
    x = P(1); y = P(2);            %           equations
    F = [ sin(x)-y^3-8;            % Equation_1
             x+y-3 ];              % Equation_2
    J = [ cos(x), -3*y^2;          % Jacobian 2 x 2
             1,      1];           %            matrix
```

(*Note* To solve three or more equations, the function M-file can be prepared similarly.) The MATLAB command is simply

```
>> Root=full_new('f_ex2n', [4.7 -1.9]')          % A
```

```
>> Root=full_new('f_ex2n', [4.7 -1.9]', tni)          % B
>> Root=full_new('f_ex2n', [4.7 -1.9]', tni,tol)      % C
>> Root=full_new('f_ex2n', [4.7 -1.9]', tni,tol, hi)  % D
```

where *tni* is the total number of iterations requested (say 12) starting from $\mathbf{x}^{(0)} = [4.7 \quad -1.9]^{\top}$, *tol* is the stopping tolerance for $|\mathbf{F}| < tol$ (say 1.0E−4) and *hi* specifies if intermediate iterates are shown ($hi = 1$ for 'yes', $hi = 0$ for 'no'). One can see that the usage is identical to the 1D case.

15.4.2 2D: Method 2 — Gauss–Jacobi

The Gauss–Jacobi iteration involves rearranging $\mathbf{F}(\mathbf{x}) = \mathbf{0}$ as $\mathbf{x} = \mathbf{G}(\mathbf{x})$ and forming the iteration

$$\mathbf{x}^{(n+1)} = \mathbf{G}(\mathbf{x}^{(n)}) \qquad \text{for } n = 0, 1, 2, \ldots.$$

This method, taking a variety of names in different contexts, is intuitively simple. However, as with 1D, the Newton–Raphson method can be considered as a special case of Gauss–Jacobi if we choose

$$\mathbf{G}(\mathbf{x}) = \mathbf{x} - J^{-1}(\mathbf{x})\mathbf{F}(\mathbf{x}).$$

For the above Example 3, $\mathbf{F}(\mathbf{x}) = \mathbf{0}$ may be rewritten as $\mathbf{x} = \mathbf{G}(\mathbf{x})$ with

$$\mathbf{G}(\mathbf{x}) = \begin{pmatrix} 3 - y \\ (\sin x - 8)^{1/3} \end{pmatrix}.$$

Here the two equations in (15.3) have been swapped; this does not change the solution.

The M-file : gauss_ja.m

The M-file `gauss_ja.m`, implementing the Gauss–Jacobi method, can be used similarly to `full_new.m`, but the main difference lies in the preparation of function M-files. As presence of the J matrix is not required, it is somewhat simpler.

Thus Example 3 above can be solved by (similar to using `full_new.m`)

```
>> Root = gauss_ja('f_ex2g', [4.7 -1.9]')           % A
>> Root = gauss_ja('f_ex2g', [4.7 -1.9]',tni,tol,hi) % D
```

where the shorter function M-file `f_ex2g.m` contains

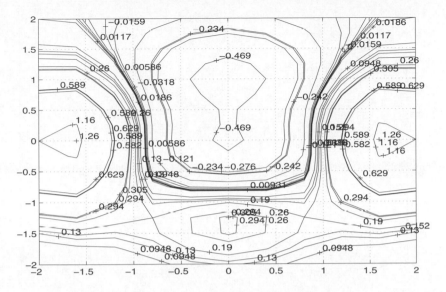

Fig. 15.4. Contour plot of $F_1(x, y)$ of two variables.

```
function G = fun_name ( P );  %  Ex 3 - (2 variables)
x = P(1); y = P(2);           %          (Gauss-Jacobi)
G = [ 3-y;                    % Here G is of size 2
     -(8-sin(x))^(1/3) ];     % Avoid complex roots
```

15.5 2D: Contour plot and convergence history

If we plot function values of **F**, the roots of **F(x)** are intersections of all zero contour lines. This gives us some rough idea about where the roots are and where we should start iterations from; refer to the M-file cont4.m.

To produce a contour plot, the main preparation involves the setting up of a uniform grid covering a suitable rectangle region. This is done via the command† meshgrid. Suppose that we want to draw a contour plot of function $F_1(x, y) = (x^6 - y^3 - 0.5) \exp(-x^2 - y^2)$ in say $\Omega = [-2, 2] \times [-2, 2]$. The procedure is the following (see Figure 15.4)

```
x0 = -2 : 0.4 : 2;
y0 = -2 : 0.25 : 2;                % Two 1D vectors
```

† For users of MATLAB V3.5, the command is meshdom.

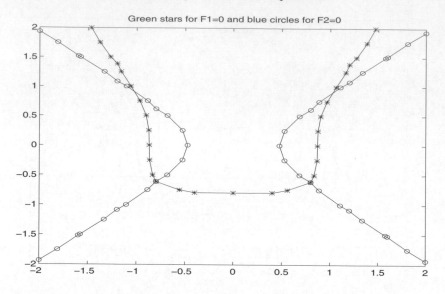

Fig. 15.5. Intersecting zero contours of functions $F_1(x, y)$ and $F_2(x, y)$.

```
[x y] = meshgrid (x0, y0) ;           % Two matrices
F1=(x.^6-y.^3-0.5) .* exp(-x.^2-y.^2); % F1 = matrix
M=max(F1);  m=min(F1);                % Row vectors
V=[m  0  M];                          % Some values of F1
C1 = contour(x,y,F1, V, '-r');        % MAIN step
clabel(C1);                           % Optional labels
grid ;                                % Optional grid
```

If M=max(max(F1)) and m=min(min(F1)) are used instead, M and m are scalars and so **V** is of size 4. In fact, **V** may be set to any desirable values within the range of **F1**.

If the zero contour lines of a second function (say) $F_2(x, y) = 20 * (x^2 - y^2) - 5 = 0$ are superimposed on those of $F_1(x, y)$ above, using different plotting symbols, the intersections must be the solution. This is also done in the (second part of) M-file cont4.m; see the results in Figure 15.5 where we can see rough locations of four roots.

Obviously a converged root (one of the four roots) of the system $F_1(x, y) = 0$ and $F_2(x, y) = 0$ should be close to a zero contour line (or an intersection of lines). Further if we plot all iterates on the contour plot, then we should see a zigzag curve converging to the root. Try the following for example; you may modify cont4.m

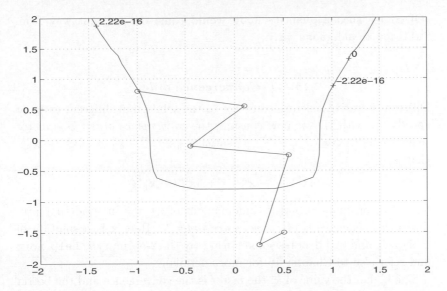

Fig. 15.6. Contour plot of $F_1(x, y) = 0$ and convergence history.

```
>> Root = [ 0.5  -1.5;    0.25  -1.7;     0.55 -0.25;
>>            -0.45 -0.1;    0.1   0.55;    -1.0  0.8];
>> x0=-2:0.4:2;    y0=-2:0.25:2;   R=Root;
>> [x y] = meshgrid (x0, y0) ;
>> F1=(x.^6-y.^3-0.5) .* exp(-x.^2-y.^2) ;   % F1=matrix
>> V=[-eps   0   eps];            % Vector of almost 0's
>> C1 = contour(x,y,F1,V, '-r');             % MAIN step
>> clabel(C1); grid ;   hold on;
>> plot(R(:,1),R(:,2),'-g',   R(:,1),R(:,2),'ow');
>> hold off
```

The plot is shown in Figure 15.6, where we assume Root contains these
iterates

```
Root = 0.5  -1.5    % x^(0) - starting vector
       0.25 -1.7
       0.55 -0.25
      -0.45 -0.1
       0.1   0.55
      -1.0   0.80 % Last iterate -- as root
```

(Of course such a quantity `Root` would come out straightaway from `full_new.m` and `gauss_ja.m`.)

15.5.1 Convergence order

Different iterative schemes converge to the solution at different rates or speeds. As with 1D, we now consider the *convergence order*. If a vector sequence $\mathbf{x}^{(0)}$, $\mathbf{x}^{(1)}$, ..., $\mathbf{x}^{(n)}$, ... satisfies $\|\mathbf{x}^{(n+1)} - \mathbf{x}\| \propto \|\mathbf{x}^{(n)} - \mathbf{x}\|^k$, that is,

$$\|\mathbf{x}^{(n+1)} - \mathbf{x}\| = C\|\mathbf{x}^{(n)} - \mathbf{x}\|^k,$$

for some constants C and k (independent of n), the method that produces this sequence is of *convergence order* k. Here $\mathbf{x}$ is a solution of $\mathbf{F}(\mathbf{x}) = 0$, and $\|\cdot\|$ denotes some norm (say the 2-norm; type `help norm` and refer also to Chapter 16 for more details).

The higher the value of k, the faster is the convergence and the better is the underlying numerical method. As with 1D, to estimate constant C and order k, we use the following M-file.

The M-file : `f_rate.m`

If the complete sequence is more than four iterates, we may use the M-file `f_rate.m` to calculate estimates of k. If `root` is the sequence of iterates that are obtained from `full_new.m` or `gauss_ga.m`, type

```
>>  k = f_rate (root, hi)
```

where $hi = 1$ asks for all estimates of k to be recorded (while $hi = 0$ only requires the averaged and final k estimate to be recorded in k).

For Example 3, we may call `f_rate.m` as follows:

```
>> Root=full_new('f_ex2n',[4.7 -1.9]', 30,1.0E-4, 1) % D
>> K = f_rate(Root, 1)
>> mean(K)
```

where `mean` finds an average value. As a final example, specify

```
>>    root = [ 3.46   2.5;
               3.5    2.59;
               3.61   2.65;
               3.735  2.69;
               3.755  2.695]
```

and call f_rate.m as follows

```
>>  K0 = f_rate(root, 0)   %% or
>>  K1 = f_rate(root, 1)
```

to obtain the results (average = 2.8728)

```
K0 = 2.8728    3.3453
K1 = 2.4003    3.3453
```

Exercises

15.1   Using full_new.m and $x^{(0)} = 1.5$, find a real root of

$$F(x) = e^x + e^{-x} + 2\cos x - 6.$$

15.2   Apply gauss_ja.m and re-solve Exercise 15.1 by selecting two different $G(x)$ functions to find the root in $[1, 2]$; e.g. $x = \ln(6 - e^{-x} - 2\cos x)$ may be a choice.

15.3   Show that if $|x^{(1)} - x| = C|x^{(0)} - x|^k$, $|x^{(2)} - x| = C|x^{(1)} - x|^k$ and all terms are assumed to be nonzero, then we have formulae

(a) $k = \ln\dfrac{|x^{(2)} - x|}{|x^{(1)} - x|} \Big/ \ln\dfrac{|x^{(1)} - x|}{|x^{(0)} - x|}$;

(b) $C = |x^{(2)} - x| / |x^{(1)} - x|^k$.

15.4   A small object, dropped from rest, encounters air resistance as it falls. Using Newton's second law to model this situation we obtain the following equation for the height H in metres that it falls in t seconds:

$$H(t) = 12.25 \left[t + 1.25(e^{-0.8t} - 1) \right].$$

To determine the time taken to fall 6.125 m, we must solve the equation $H(t) = 6.125$, that is,

$$F(t) = 1.25e^{-0.8t} + t - 1.75 = 0.$$

Using the M-file gauss_ja.m for each of the following methods:

(a) Newton–Raphson: $G_0(t) = t - F(t)/F'(t)$ (as in §15.2);
(b) Gauss–Jacobi: use $G_1(t) = 1.75 - 1.25e^{-0.8t}$;
(c) Gauss–Jacobi: use $G_2(t) = t - \dfrac{F(t)}{\sqrt{1 - (0.8t + 0.6)\exp(-0.8t)}}$

and with $t^{(0)} = 0.5$,

(i) determine a solution correct to four decimal places;

(ii) estimate the convergence order in (i) if possible;

(iii) find and plot the convergence region.

Hint Use the search region $[-5, 5]$ and `axis([-5 5 -4 4])` after `plot`.

15.5   Given the nonlinear function $F(x) = 0.6 - (3 + \ln x)x + e^x$:

(a) Generate four random numbers x_1, x_2, x_3 and x_4 in intervals $(1, 1.3)$, $(1.4, 1.5)$, $(1.5, 1.6)$ and $(1.7, 2)$ respectively, and evaluate $F(x)$ at $x = x_1, x_2, x_3, x_4$. For each root of $F(x)$ in $[1, 2]$, find an interval that contains it. Mark all four points on a graph of the function $F(x)$ over some search region and verify that it has exactly two roots α and β in $[1, 2]$.

Hint Use the intermediate value theorem on sign changes of a function.

(b) Try to use the Gauss–Jacobi method $(tol = 10^{-5})$ with $G(x) = 0.6 - (2 + \ln x)x + e^x$, to compute roots α and β and estimate the convergence order for each case; if this $G(x)$ is not suitable for β, you need to find something that is suitable.

15.6   Using the Taylor theorem in one variable, deduce that, for a single function $H = H(\mathbf{x}) = H(x_1, x_2)$ in two variables $(m = 2)$,

$$H(\mathbf{x}) \quad = \quad H(\mathbf{x}^{(0)}) + (\mathbf{x}^{(1)} - \mathbf{x}^{(0)}) \cdot \nabla H + hot,$$

where $\nabla H = \mathrm{grad} H = (\frac{\partial H}{\partial x_1}, \frac{\partial H}{\partial x_2})$. Use this formula to find a linear approximation to

$$H(\mathbf{x}) = \ln(x_1^5 x_2^6) \qquad \text{at } \mathbf{x}^{(0)} = (1, \ 1)^\top.$$

EXTRAS Can you generalise this result to the case of three variables $(m = 3)$ and give an example?

15.7   Consider the system $\mathbf{F}(\mathbf{x}) = \mathbf{0}$, given by

$$\begin{cases} F_1(x_1, x_2) = e^{x_1 - x_2} - \sin(x_1 + x_2) = 0, \\ F_2(x_1, x_2) = x_1^2 x_2^2 - \cos(x_1 + x_2) = 0. \end{cases}$$

(a) Using `full_new.m` and $\mathbf{x}^{(0)} = (-2, 1)^\top$, find a root up to four digits of accuracy.

(b) Using `gauss_ja.m` and $\mathbf{x}^{(0)} = (0.5, 0)^\top$, find a root with no more than 40 iterations using the formula $\mathbf{G} = \mathbf{x} - \omega J \backslash \mathbf{F}$. Take $\omega = 0.5$, and $\omega = 1.5$.

Hint When $\omega = 1$ the method is identical to the Newton method. Can you find a different $\mathbf{G}(\mathbf{x})$ formula that works?

15.8 Show that if $\|\mathbf{x}^{(1)}-\mathbf{x}\|/\|\mathbf{x}^{(0)}-\mathbf{x}\|^k = \|\mathbf{x}^{(2)}-\mathbf{x}\|/\|\mathbf{x}^{(1)}-\mathbf{x}\|^k = C$, and all terms are assumed to be nonzero, we have

(a) $k = \ln \dfrac{\|\mathbf{x}^{(2)} - \mathbf{x}\|}{\|\mathbf{x}^{(1)} - \mathbf{x}\|} \Big/ \ln \dfrac{\|\mathbf{x}^{(1)} - \mathbf{x}\|}{\|\mathbf{x}^{(0)} - \mathbf{x}\|}$;

(b) $C = \|\mathbf{x}^{(2)} - \mathbf{x}\|/\|\mathbf{x}^{(1)} - \mathbf{x}\|^k$.

15.9 Following Exercise 15.7,

(a) using a grid spacing of 0.2 on the region

$$R = \{\mathbf{x} : -2 \le x_1 \le 1.6,\ 0 \le x_2 \le 2\}$$

deduce, from the contour plots of F_1 and F_2 on the same graph, that there are two roots in R;

(b) for the root nearest to the point $(0.5, 0.0)^\top$, plot its Newton iterates on the contour plot and find the averaged convergence rate of the iterates. Further plot on the same graph Newton iterates starting from $(-2, 1)^\top$.

15.10 Consider the system of equations $\mathbf{F}(\mathbf{x}) - \mathbf{0}$ given by

$$\begin{cases} F_1(\mathbf{x}) = 12x_1 - 3x_2^2 - 4x_3 - 7.17 = 0, \\ F_2(\mathbf{x}) = x_1^2 + 10x_2 - x_3 - 11.54 = 0, \\ F_3(\mathbf{x}) = x_2^3 + 7x_3 - 7.631 = 0. \end{cases}$$

(a) Write down the Jacobian matrix for $\mathbf{F}$ and use the Newton–Raphson method with $\mathbf{x}^{(0)} = (0,\ 0,\ 0)^\top$ to obtain the solution correct to five decimal places.

(b) Use the third equation to eliminate x_3 and use Newton–Raphson to solve the reduced system of two equations for x_1 and x_2 correct to five decimal places. Evaluate x_3.

(c) **EXTRAS** Using a grid spacing of 0.1 on the region

$$R = \{\mathbf{x} : -5 \le x_1 \le 25, -15 \le x_2 \le 5\},$$

obtain the contour plots of F_1 and F_2, based on (b), and further decide on the total number of solutions to the original system. Find any remaining solutions correct to five decimal places using Newton–Raphson. Write down a version of Gauss–Jacobi iteration formulae for the original system and investigate the convergence of this method to each root.

(d) **EXTRAS** Other methods exist for the solution of nonlinear equations but we leave you to find out about the
MATLAB command `fsolve` which† is based on a method
of nonlinear least-squares. Compare its performance with
`full_new.m`.

---

† For some earlier versions of MATLAB (including V.4.0), this command is not
available.

16

Matrices and Solution of Linear Systems

Linear systems of algebraic equations are one of the most important subjects in mathematics, since most other subjects, methods and problems involve or reduce to this subject.

Aims of the project

As numerical methods involve truncations and finite precisions, we investigate their effect on solution accuracy. Most of us know some theory about linear systems but may not be aware of the good or bad choice of solution methods on computers, what determines the accuracy of the numerical solution and whether an obtained solution can be improved. Large scale problems arising from practice often involve sparse matrices and special techniques can be developed. This project addresses all such issues.

Mathematical ideas used

You have learnt that to solve $A\mathbf{x} = \mathbf{b}$ you just type $\mathbf{x} = A\backslash\mathbf{b}$. To investigate the sensitivity of the solution $\mathbf{x}$ with regard to the matrix A (or its condition number) and the right-hand side vector $\mathbf{b}$, we use a controllable number of digits in our calculations. The use of iterative refinements is illustrated. Finally we discuss how a sparse matrix may be condensed towards band forms by use of permutation matrices.

MATLAB techniques used

Seven M-files lin_solv.m, chop.m, lu2.m, lu3.m, lu4.m, solv6.m and spar_ex.m are used to assist this project. The last two M-files are listed in the chapter. Here chop.m, used to fix digits, is used by lu2.m, lu3.m and lu4.m. You will get useful experience of MATLAB's easy and simple commands for both dense and sparse matrices.

16.1 Operation counts

Operation counts give an indication of how many floating point operations (flops) are needed for a given mathematical problem. Any reduction in operation counts leads to a faster solution. This project will highlight three basic ideas to this end. The first idea, discussed below, is to implement an existing method in a more efficient way in order to achieve speed-up. The second idea, discussed in §16.2 and §16.3, is to seek new and alternative methods for solving the same problem. The third idea, discussed in §16.5, is to reformulate the given problem so that existing methods can be efficient.

To illustrate the first idea, we consider the simple task of calculating $\mathbf{z} = ABC\mathbf{b}$, where A, B, C are $n \times n$ matrices, and $\mathbf{b}$ is a vector of size n. A naive way to do this is to work out a matrix $D = ABC$ first before doing $\mathbf{z} = C\mathbf{b}$; this costs roughly $4n^3$ floating point operations (flops). A better method is to compute three matrix-vector products, that is, $\mathbf{z}_1 = C\mathbf{b}$, $\mathbf{z}_2 = B\mathbf{z}_1$, $\mathbf{z} = A\mathbf{z}_2$; this costs only $6n^2$ flops!

The efficiency of different methods is therefore crucially determined by operation counts. We may get estimates of operation counts theoretically from the mathematical formulae but such a job can be done easily by the MATLAB command `flops`; to demonstrate its usage, consider the second method in the above example

```
>>  flops(0); n=2000;      % Set flops counter to zero
>>  A=rand(n); B=rand(n);  % Setup matrices
>>  C=rand(n);b=rand(n,1); % Setup matrices
>>  z=C*b; z=B*z; z=A*z;   % Operations done
>>  w=flops, ratio=w/n^2   % Work out flops
```

where you should find $w = 24000000$ and $ratio = 6$. Note that `rand` does not contribute to `flops`.

Apart from `flops`, three other commands can be used to monitor the time spent by MATLAB to carry out a particular calculation; these are shown in Table 16.1, where 'Method' refers to operations of a numerical method and row 'E' shows results from doing the simple operation `rand(9)*5`, that is, replacing 'Method' by `rand(9)*5`. Note `etime` checks the difference in `clock` and converts it into seconds; type `help etime` to be aware of a possible problem.

Remark Broadly speaking many classical mathematical methods, although elegant in theory and useful in proving the existence of solutions, are not suitable for computer use. For they may either take too

Table 16.1. *Monitoring the efficiency of a numerical method.*

flops	tic/toc	cputime	etime
>> flops(0)	>> tic	>> t0=cputime;	>> t0=clock;
>> 'Method'	>> 'Method'	>> 'Method'	>> 'Method'
>> t=flops	>> t=toc	>> t=cputime; >> t=t-t0	>> t=clock; >> t=etime(t,t0)
E t = 81	t = 6.09	t = 5.27	t = 3.13

long (even years) or lead to large errors; refer to §16.2.3. For example, $A^{-1} = adj(A)/det(A)$ is good for a theory but not suitable for numerical computations because the number of operations required is too large! Here $adj(A)$ denotes the adjoint of matrix A — a matrix of cofactors. Therefore the analysis, selection (validation) and development of practical and useful numerical methods are the main aims of computational mathematics.

16.2 Dense linear systems

In what follows, you may issue the MATLAB command format compact first to control output.

16.2.1 *MATLAB solution of* $Ax = b$ *and pivoting*

For square matrices, the simple instruction $\mathbf{x} = A\backslash\mathbf{b}$ is not the same as $\mathbf{x} = \text{inv}(A) * \mathbf{b}$ because each represents a different method for solving the same problem. Convince yourself by doing the following MATLAB experiment before proceeding. First generate a random matrix A of size 8×8 using A=rand(8) and a column vector $\mathbf{b}$ of size 8 using b=rand(8,1). Then type

```
>>    flops(0) ;      % Set flops counter to zero
>>    x = A\b         % A is 8 x 8  and b is 8 x 1.
>>    f1 = flops      % Give number of operations used
>>    flops(0) ;      % Set counter to zero again
>>    x = inv(A)*b
>>    f2 = flops      % Should be larger
```

Here and from now on where `flops` is used, it is always possible to use the other MATLAB commands listed in Table 16.1.

The actual MATLAB method for solving $Ax = \mathbf{b}$ is based on the Gauss elimination method and may be described as a three-stage approach:

(1) Compute a lower triangular matrix L, an upper triangular matrix U and a permutation matrix P such that $PA = LU$. Here P is simply the identity matrix I with some of its rows swapped. This swapping ensures that in reducing the new matrix $A_1 = PA$ to its row echelon form, *all multipliers in the elimination process are less than or equal to one.* The use of P implies that we solve $PA\mathbf{x} = P\mathbf{b}$ or rather $LU\mathbf{x} = P\mathbf{b}$ instead of $A\mathbf{x} = \mathbf{b}$. Such a process is known as *partial pivoting* and helps to maintain the accuracy of computations.

(2) Solve $L\mathbf{y} = P\mathbf{b}$, known as forward substitution.

(3) Solve $U\mathbf{x} = \mathbf{y}$, known as backward substitution.

The first stage known as *triangular factorisation* (or LU decomposition), related to the row echelon form reduction,† is the most important step. It can be done via the command `lu`. Therefore the following‡ is equivalent to using $\mathbf{x} = A\backslash\mathbf{b}$

```
>> [L,U,P] = lu(A); % Find L, U and P matrices (A1=PA=LU)
>> B = P*b ;        % A x = b  becomes  A1 x=B or LU x=B
>> y = L \ B ;      % Solve  L y = B    as  L (U x) = B
>> x = U \ y        % Solve  U x = y              ===
```

16.2.2 *L and U are reusable*

Suppose that we hope to solve $A\mathbf{x} = \mathbf{b}$ and $A\mathbf{x} = \mathbf{e}$ (or for that matter a number of linear systems with the same matrix but different right-hand sides). Using the $\backslash$ approach twice is *not* a good idea, because each time MATLAB would do one LU factorisation of PA.

A better procedure is the following:

---

† The MATLAB command `rref` does this job but one cannot see much output from it. Another command `rrefmovie` is more useful. Fortunately one is allowed to view the sources of `rref.m` and `rrefmovie.m`. Try `A=round(rand(6)*7); [R J]=rref(A)`, `rrefmovie(A)`.

‡ Note that $y = $ `L\B` does not take as many operations as `A\b` because MATLAB is clever enough to recognize the triangular nature of L! The same is true for $x = $ `U\y`. Test this out using `flops`.

```
>> [L,U,P] = lu(A);       % Find L,U,P matrices (PA=LU)
>>           B = P*b;      %        Solve  PAx=Pb  (eqn 1)
>>           z = L \ B;    %        Solve  PAx=Pb
>>           x = U \ z     %        Solve  PAx=Pb
>>           E = P*e;      %        (eqn 2)   Solve PAx=Pe
>>           z = L \ E;    %                  Solve PAx=Pe
>>           x = U \ z     %                  Solve PAx=Pe
```

16.2.3 Significant decimal digits

. In MATLAB and most other computer packages or languages, only
a finite number of digits are actually used in calculations, and so most
numbers must be rounded. This gives rise to computing errors. The
more computing steps we have, the larger the final error. This partially
explains why we should always abandon a method like $x = inv(A) * b$
on computers, as it not only takes much longer for a start but also may
accumulate larger errors.

All MATLAB calculations by default use 16 decimal digits. As we
want to investigate the effect of finite precision arithmetic on numerical
accuracy, we shall mostly work with t-digit arithmetic in this project
with $1 \le t \le 16$. The M-file chop.m provides us with an easy way to do
this. Try help chop, chop(0.1364, 2) * 10 or chop(1.223345, 6)
and chop(1.223345, 5) to see how it works.

16.2.4 The M-files: lin_solv.m, lu2.m and lu3.m

We shall explore the importance of pivoting and significant digits in the
context of solving $Ax = b$. The main M-file lin_solv.m is used as
follows

```
>>        x = lin_solv( 'lu_name', A, b, t) ;
```

where lu_name is the name of the factorisation function M-file which may
be replaced by lu2 or lu3 or lu4. Here t indicates that all calculations
are performed to the fixed precision of t-digits only.

The function M-file lu2.m is mathematically the same as the built-in
system file lu.m except that now we only perform t-digit arithmetic.
So if you input $t = 16$ with lu2.m, then lin_solv.m should produce
identical results to $x = A\backslash b$.

The function M-file lu3.m does not allow partial pivoting and is only
to be used as a comparison, not to compute accurate solutions as no

permutation matrix P is involved and multipliers larger than one are permitted. The M-file `lu4.m` always produces the same results as `lu3.m` but uses fewer flops for sparse matrices; see §16.5.

The two M-files `lu2.m` and `lu3.m` can be used individually for factorisation (similar to `lu`)

```
>> [L,U,P]=lu2(A,t); %Factorise A (t digits &    P)
>> [L,U]  =lu3(A,t); %Factorise A (t digits & no P)
```

So if you solve $A\mathbf{x} = \mathbf{b}$ by

```
>>        x = lin_solv( 'lu2', A, b, t) ;
```

we may say the method is Gaussian elimination with partial pivoting. Similarly the following is called Gaussian elimination without partial pivoting

```
>>        x = lin_solv( 'lu3', A, b, t) ;
```

16.3 The iterative refinement algorithm

If $\mathbf{y}$ is an approximate solution of $A\mathbf{x} = \mathbf{b}$, define its *residual* as $\mathbf{r} = \mathbf{b} - A\mathbf{y}$ which may not be zero (or small). If we solve $A\mathbf{e} = \mathbf{r}$, then we can usually expect $\mathbf{x}_1 = \mathbf{y} + \mathbf{e}$ to be a better approximation to the true solution $\mathbf{x}$. The procedure may be repeated if the residual for $\mathbf{x}_1$ is not 'small' enough and we have an iterative process. This is the basic idea of the *iterative refinement algorithm*† used to improve solutions:

Step 0. Use Gauss elimination with a fixed precision $t < 16$ to solve $A\mathbf{x} = \mathbf{b}$ and keep L, U, P matrices. Denote the approximate solution by $\mathbf{y}$. (*Note $P = I$ when no partial pivoting is used.*)

Step 1. Compute the residual $\mathbf{r} = \mathbf{b} - A\mathbf{y}$ (if $\mathbf{r} = \mathbf{0}$ or rather if its norm‡ is small enough, we stop here and accept $\mathbf{x} = \mathbf{y}$).

Step 2. Use the same L, U, P to solve $LU\mathbf{e} = P\mathbf{r}$.

Step 3. Update the solution $\mathbf{y} \leftarrow \mathbf{y} + \mathbf{e}$.

Step 4. Go back to Step 1 and repeat.

---

† We remark that iterative refinement is closely related to another idea called *residual correction*, where A may represent other operators (e.g. differential, integral, nonlinear). This idea forms the basis of a modern and powerful numerical technique — the multi-level method (or multi-grid method).

‡ With MATLAB, if $\mathbf{r} \in \mathcal{R}^n$, then its p-norm is $\|r\|_p = \text{norm}(\mathbf{r}, \mathbf{p})$ with p being either any positive integer or 'inf' (i.e. for ∞-norm). However, for matrices, MATLAB command norm can only compute four norms, namely, $\|A\| = \text{norm}(\mathbf{A}, \mathbf{p})$, where p can be 1, 2, 'inf' or 'fro' that denotes, respectively, 1-norm, 2-norm, ∞-norm and F-norm (Frobenius). The default norm(A) gives the 2-norm.

The values in both $\mathbf{r}$ and $\mathbf{e}$ give indications of the accuracy of $\mathbf{y}$. They can be measured by the MATLAB command `norm`.

16.4 A perturbation analysis for $Ax = b$

In general, the accuracy of $\mathbf{x} = A\backslash\mathbf{b}$ depends on the property of matrix A (the so-called *conditioning*). A measure of *conditioning* is by its condition number which is defined by $cond(A) = \|A^{-1}\|\,\|A\|$, where $\|\cdot\|$ can be any matrix norm. However, for the condition number, MATLAB has a simple command† `cond` in the 2-norm; just type `cond(A)`. For 'good' matrices, the digit number t does not affect accuracy much; but for 'bad' matrices‡ (or *ill conditioned* ones), t is significant!

Now consider the solution accuracy when $t < 16$. Suppose $Ax = \mathbf{b}$ denotes the true linear system involving no rounding in A and $\mathbf{b}$. However, in reality, due to rounding and possible data collection error, our numerical solution $\mathbf{y}$ actually only satisfies a so-called perturbed system (nearby)

$$(A + \Delta A)\mathbf{y} = \mathbf{b} + \Delta\mathbf{b}.$$

Here $(A + \Delta A)$ is viewed as the matrix A in t precision and similarly for $\mathbf{b} + \Delta\mathbf{b}$. The error terms ΔA (matrix) and $\Delta\mathbf{b}$ (vector) are usually known. For example, when we use t-digit arithmetic,

$$\|\Delta A\| = \texttt{norm(A - A1)} \quad \text{and} \quad \|\Delta\mathbf{b}\| = \texttt{norm(b - b1)},$$

where $A1 = \texttt{chop}(A, t)$ and $\mathbf{b}1 = \texttt{chop}(\mathbf{b}, \mathbf{t})$.

The solution accuracy is a measure of how close the numerical solution $\mathbf{y}$ is to the true solution $\mathbf{x}$. A linear system is *well conditioned* if small values of $\|\Delta A\|/\|A\|$ and $\|\Delta\mathbf{b}\|/\|\mathbf{b}\|$ lead to a small value of $\|\mathbf{x}-\mathbf{y}\|/\|\mathbf{x}\|$. Otherwise, it is *ill conditioned*. Obviously we want to know how these quantities are related. We have the following result:

Theorem 16.1 *Suppose that the exact equation is* $Ax = \mathbf{b}$ *and we have actually solved*

$$(A + \Delta A)\mathbf{y} = \mathbf{b} + \Delta\mathbf{b}.$$

The relative error will be

---

† To compute $cond(A)$ in other norms, we may use the definition involving A^{-1} but this can be expensive in terms of `flops`. For 1-norm, try `condest(A)`.

‡ For a 'bad' matrix, to achieve a good accuracy, the solution strategy is to convert it into a 'good' one rather than to demand an unlimited increase of t. Such a method is called *pre-conditioning*.

$$\frac{\|\mathbf{x} - \mathbf{y}\|}{\|\mathbf{x}\|} \le C \left(\frac{\|\Delta A\|}{\|A\|} + \frac{\|\Delta \mathbf{b}\|}{\|\mathbf{b}\|} \right),$$

where the constant is given by

$$C = \frac{cond(A)}{1 - cond(A) \left(\frac{\|\Delta A\|}{\|A\|} \right)}.$$

Therefore we say a problem $A\mathbf{x} = \mathbf{b}$ is *well conditioned* if C is not too large and *ill conditioned* if C is large. For the purpose of this project, let us call a problem *ill conditioned* if $C \ge 50$.

Example Solve a random 11×11 linear system $A\mathbf{x} = \mathbf{b}$ with $t = 2$ digit arithmetic; see Chapter 6 for using `rand`. We want to know if the problem is well conditioned. The following M-file is also available as `solv6.m`

```
rand('seed',1998); n=11;        % To fix 'seed'
A = 10*rand(n,n) ;
b = rand(n,1);                  % Generate the system
t = 2 ;                         % Set the precision
x = A \ b ;                     % 'Exact' solution
y = lin_solv( 'lu3', A, b, t) ; % Numerical solution
A_rel=norm(A-chop(A,t))/norm(A);% Relative error in A
b_rel=norm(b-chop(b,t))/norm(b);% Relative error in b
k = cond(A) ;                   % Condition number
C = k / ( 1 - k * A_rel ) ;
if C >= 50, disp('Problem is ill conditioned ...'),
else,       disp('Problem is well conditioned.'), end
error_theory = C*(A_rel+b_rel)  % Formula in Theorem
error_found  = norm(x-y)/norm(x)% Exact x, numerical y
```

Here you should find that $C = 97.561$ so the problem is ill conditioned, `error_theory` $= 0.7618$ is the predicted error by theory (larger) and `error_found`$=0.3381$ is the actual error observed (smaller).

16.5 Sparse matrices, graph ordering and permutations

Matrices that are generated from applied mathematical problems are often sparse; by 'sparse' we mean many entries (say 50%) are actually zeros. Therefore avoiding any operations with these zero positions can lead to the speeding up of practical algorithms. One general approach is to use permutation matrices to gather all nonzeros together to form a desirable pattern (e.g. band matrices that have zeros in all entries except on the main diagonal and a small number of subdiagonals).

To visualise sparse matrices, we may use the MATLAB command `spy`, for example, try the following (refer to M-file `spar_ex.m`)

```
>> A = diag(0:9);      % Generate a matrix of size 10
>> spy(A); grid        % See what it looks like
>> a = eye(10,1)*ones(1,10)*2;  % Nonzero Row 1
>> b = ones(10,1)*eye(1,10)*3;  % Nonzero Column 1
>> A = A + a + b;      % Formed an arrow head matrix
>> spy(A); grid        % See what it looks like now
```

To appreciate the techniques used in sparse matrices, consider the following

```
>> A1 = A;             % Note this A follows the above
>> flops(0)            % Counter reset (try also tic/toc)
>> [L1 U1]=lu3(A1);%   Factorise A as a full matrix
>> Work_1 = flops      % Operations performed
>> a1 = symrcm(A1)     % Find new ordering for matrix A1
>> A2 = A1(a1,a1)      % Reorder A2=P*A1*P'
>> flops(0)            % Counter reset (try also tic/toc)
>> [L2 U2]=lu4(A2);%   Factorise A as a sparse matrix
>> Work_2 = flops      % Operations performed (new)
```

You should find that Work_1 = 8434 and Work_2 = 1506. Here the trick is that we have found a new ordering for matrix $A1$; to see and use the permutation matrix, try

```
>> P = eye(10);        % Generate a 10 x 10 identity matrix
>> P = P(a1,:)         % Reorder its rows
>> A3 = P*A1*P'        % Here A3 = A2 !
```

As we know, P is orthogonal† and thus the eigenvalues of A, $A2$ and $A3$ are the same. Evidently it is more advantageous to work with $A3$ to find eigenvalues.

Finally, we show how to represent sparse matrices by graphs! A sparse matrix $A_{n \times n}$ can be viewed as a representation of zeros and nonzeros, and each nonzero represents a relationship (or an interaction) between

---

† Check this out by verifying $P^\top P = I$. *Note:* Advanced users of MATLAB may use `speye`, instead of `eye`, to form an identity matrix; try also `help sparse` to find out more about building sparse matrices.

two numbers (i.e. a row and a column index, both belonging to the same index set 1, 2, ..., n). In graph theory, a graph G with n nodes (not all nodes connecting each other) may have a variable number of edges, each edge representing a connection of two nodes.

Therefore, a sparse matrix A can be naturally linked with a graph $G(A)$ if we 'identify' an index from the former with a node from the latter, and also a nonzero from the former with an edge with the latter. Obviously zero entries of A correspond to nonexistent edges of $G(A)$! Thus the connectivities (edges) of n nodes of a graph can represent a sparse matrix. More precisely, edges of a graph $G(A)$ represent nonzero elements of A, that is to say, node i connects node j in the graph if and only if entry $A_{ij} \neq 0$ in the matrix.

For example, a tridiagonal matrix can represent n nodes forming an open chain (connectivity), with each middle node only connecting its two immediate neighbours, that is, the first end node 1 only connecting to node 2, the middle node k only connecting to nodes $k-1$ and $k+1$ and the last end node n only connecting to node $n-1$. To put it the other way round, with such a $G(A)$, node k only connects to nodes $j = k-1$ and $j = k+1$, and so for the underlying matrix, entry $A_{kj} \neq 0$ only for $|k - j| \leq 1$. This implies that A must be tridiagonal!

In the above example of a 10×10 matrix, the original matrix A represents the connectivities of 10 nodes in the natural ordering $\mathbf{a} = 1 : 10$ and $A2$ represents the connectivities in a new ordering $\mathbf{a1} = [10\ 9\ 8\ 4\ 6\ 5\ 2\ 7\ 1\ 3]$.

For a symmetric matrix A, its graph $G(A)$ can be plotted by the MATLAB command `gplot`. For the above example with $A_{10\times10}$, try the following (available in the M-file `spar_ex.m`)

```
>> xy = [    0         0
           1.5       1.3
           0.35      2.0
          -1.0       1.7
          -1.9       0.68
          -1.9      -0.68
          -1.0      -1.7
           0.35     -2.0
           1.5      -1.3
           2.0       0.0 ],  figure(2)
>> gplot(A,xy); hold on          % Plot all edges
>> gplot(A,xy,'o')               % Plot all nodes as 'o'
```

```
>> for k=1:10,
   tt = sprintf('%d\n',k);
   text(xy(k,1),xy(k,2),tt)      % Print numbers
   end; axis off
>> title('A with the original ordering a')
>> hold off;  figure(3)          % Now plot G(A2) below
>> xy2 = xy(a1,:);               % Reorder nodes by a1
>> gplot(A2,xy2); hold on        % Plot all edges
>> gplot(A2,xy2,'o')             % Plot all nodes as 'o'
>> for k=1:10,
   tt = sprintf('%d\n',k);
   text(xy2(k,1),xy2(k,2),tt)    % Print numbers
   end; hold off; axis off
>> title('A1 with the new ordering a1')
```

The study of sparse matrix techniques is of importance in applied sciences. Interested readers may consult two books on sparse matrices, [4] and [15].

Exercises

16.1   Generate at least three random matrices $A_{n \times n}$ of random orders $n = 50$ to $n = 170$ and random column vectors b of size n. Then use **flops** to show the following results hold approximately and estimate constant C in each case:

- $x = $ A\b takes Cn^3 operations;
- $y = $ det(A) takes Cn^3 operations;
- $z = $ inv(A) takes Cn^3 operations;
- $t = $ b $*$ b$'$ $*$ b takes Cn^2 operations;
- $u = $ b $*$ (b$'$ $*$ b) takes Cn operations;
- $v = $ b $*$ 9 + 6 takes Cn operations;
- $w = $ A $*$ b takes Cn^2 operations.

16.2   Solve the following system $A\mathbf{x} = \mathbf{b}$ using both A\$\mathbf{b}$ and two separate steps of the LU approach (are the two solutions identical?)

$$
\begin{bmatrix}
1 & 1 & 0 & 3 \\
2 & 1 & -1 & 1 \\
3 & -1 & -1 & 2 \\
-1 & 2 & 3 & -1
\end{bmatrix}
\begin{pmatrix}
x_1 \\
x_2 \\
x_3 \\
x_4
\end{pmatrix}
=
\begin{pmatrix}
4 \\
1 \\
-3 \\
4
\end{pmatrix}.
$$

16.3   Try the approach of §16.2.2 to solve new systems $A\mathbf{y} = \mathbf{e}$ and
       $A\mathbf{x} = \mathbf{f}$, immediately after solving $A\mathbf{x} = \mathbf{b}$ in Exercise 16.2,
       with $\mathbf{e} = (4 \ \ 2 \ \ -2 \ \ 1)^\top$ and $\mathbf{f} = (34 \ \ 10 \ \ 1 \ \ 36)^\top$.

16.4   Based on [L U P]=lu(A) for Exercise 16.2, calculate the follow-
       ing relative errors

$$e_\mathbf{L} = \frac{\|L - L2\|}{\|L\|} \text{ and } e_\mathbf{U} = \frac{\|U - U2\|}{\|U\|},$$

       where $L2$, $U2$ are chopped from L, U with $t = 1$, and $\| \cdot \|$ here
       refers to a norm computed by the MATLAB command norm,
       for example, $\|L\|$ computed by norm(L).

16.5   Using lin_solv.m with lu2.m and lu3.m, solve Exercise 16.2
       again by $t = 1$ digit arithmetic. Compare the two solutions.
       Which one is more accurate? Use command norm as in Exercise
       16.4.

16.6   Consider the system of linear equations $A\mathbf{x} = \mathbf{b}$, where

$$A = \begin{pmatrix} 1.1756 & 4.0231 & -2.14170 & 5.1967 \\ -4.0231 & 1.0002 & 4.5005 & 1.1973 \\ -10.179 & -5.2107 & 1.1022 & 0.10034 \\ 886.19 & 7.0005 & -6.6932 & -4.1561 \end{pmatrix}$$

       and $\mathbf{b} = (\ 15.721 \ \ 19.392 \ \ 2.9507 \ \ -38.089 \ )^\top$. Take $\mathbf{xe} =$
       $A\backslash\mathbf{b}$ as the true solution.

       (a) Find the approximate solution $\mathbf{y}$ to $A\mathbf{x} = \mathbf{b}$ with $t = 4$
           decimal digit arithmetic by means of Gauss elimination
           without partial pivoting. Calculate the relative error

$$er = \frac{\|\mathbf{xe} - \mathbf{y}\|}{\|\mathbf{xe}\|},$$

           where $\| \cdot \|$ denotes the 2-norm.
       (b) Perform one iterative refinement based on (a), and cal-
           culate the new relative error.
       (c) Following (b) and using $t = 4$ again, how many more
           steps of iterative refinements do we need to reach the
           limit of iterative refinement, that is, to obtain $\|\mathbf{e}\| < 10^{-9}$
           in $\mathbf{y} \leftarrow \mathbf{y} + \mathbf{e}$ of the iterative refinement algorithm? Show
           all intermediate residuals, their norms and the relative
           error of the final solution.
           Hint use the command while as in

```
                    y = lin_solv( ... );
                    count = 0
            while ( norm(e) > 1.0E-9 )
                    r = b - A * y
                    e = . . . . . . . ;
                    y = y + e ;
                    count = count + 1
                    norm_e_is = norm(e)
            end
```

16.7   Following Exercise 16.6, solve the two modified problems below
by Gauss elimination with partial pivoting using $t = 5$ digit
arithmetic. Decide on the conditioning of each case.

   (a) Keep the original A as above but modify **b** by values that
   are randomly distributed on the interval $[-0.001, 0.001]$.
   Hint $\mathbf{b} \leftarrow \mathbf{b} + 0.002 * \mathbf{rand}(4,1) - 0.001 * \mathbf{ones}(4,1)$.

   (b) Keep the original **b** but modify A by values that are
   randomly distributed on the interval $[-0.002, 0.002]$.

16.8   **EXTRAS** Find out about how to generate a Hilbert matrix by
`help hilb`. Let A be a Hilbert matrix of order 6 and $\mathbf{b}^\top =$
$2 : 1 : 7$. Find the actual relative error E for the numerical
solution **y** by solving $A\mathbf{x} = \mathbf{b}$ by Gauss elimination without
partial pivoting and using $t = 8$ digit arithmetic.

   Modify your M-file so that it runs for $\mathbf{t} = 1 : 12$, and for each
case print out a relative error for **y**.

16.9   For the following 9×9 sparse matrix

$$
A = \begin{bmatrix}
1 & 0 & 1 & 1 & 0 & 0 & 0 & 1 & 1 \\
0 & 1 & 1 & 0 & 0 & 0 & 0 & 0 & 0 \\
1 & 1 & 1 & 0 & 0 & 0 & 0 & 0 & 0 \\
1 & 0 & 0 & 1 & 1 & 0 & 0 & 0 & 0 \\
0 & 0 & 0 & 1 & 1 & 0 & 0 & 0 & 0 \\
0 & 0 & 0 & 0 & 0 & 1 & 0 & 0 & 1 \\
0 & 0 & 0 & 0 & 0 & 0 & 1 & 1 & 0 \\
1 & 0 & 0 & 0 & 0 & 0 & 1 & 1 & 0 \\
1 & 0 & 0 & 0 & 0 & 1 & 0 & 0 & 1
\end{bmatrix},
$$

use `symrcm` to obtain a new ordering **a** and a new permuted
matrix $A_1 = \mathbf{A}(\mathbf{a}, \mathbf{a})$. Plot both graphs $G(A)$ and $G(A_1)$ using

gplot, assuming that graph $G(A)$ has these nodal positions

$$xy = \begin{bmatrix} 0 & 0 \\ 0 & 1 \\ 0 & 0.5 \\ 0 & -0.5 \\ 0 & -1 \\ -1 & 0 \\ 1 & 0 \\ 0.5 & 0 \\ -0.5 & 0 \end{bmatrix}.$$

16.10 For the following 11×11 matrix A, decompose it by lu3.m and lu4.m and find the number of flops required in each case

$$\begin{bmatrix} \mu & -2 & 0 & 30 & 3 & 3 & 0 & 4 & 0 & 3 & -2 \\ 0 & \mu & 4 & 4 & 0 & 3 & -2 & 0 & 0 & 3 & 0 \\ 0 & 4 & \mu & 3 & 0 & 0 & -2 & 0 & 0 & 0 & 0 \\ 0 & 0 & 2 & \mu & 0 & 0 & 6 & 0 & 0 & 0 & 0 \\ -2 & 6 & 0 & 0 & \mu & -2 & 4 & -2 & 0 & 4 & 0 \\ 3 & 3 & 6 & 0 & 4 & \mu & -2 & 0 & 0 & -2 & 0 \\ 0 & 2 & 4 & 4 & 0 & 0 & \mu & 0 & 0 & 4 & 0 \\ -2 & 0 & 0 & 0 & 3 & 3 & 0 & \mu & 4 & -2 & 3 \\ 2 & 0 & 0 & 0 & 3 & 0 & 0 & -2 & \mu & 0 & 6 \\ 0 & -2 & 3 & 4 & 3 & -2 & 4 & 0 & 0 & \mu & 0 \\ 4 & 0 & 0 & 0 & -2 & 2 & 0 & -2 & -2 & 0 & \mu \end{bmatrix},$$

where $\mu = 999$. Further use the MATLAB reordering† command symrcm to obtain a new matrix A_1. Compare the flops needed to decompose matrix A_1 using lu4.m with the previous two cases and work out percentages of saving. Plot all nonzero positions of A and A_1 using spy. Can you suggest an ordering that is better than symrcm?

---

† Note that this matrix is not really symmetric but symrcm can still be useful.

17

Function Interpolations and Approximation

Function approximation lies at the very heart of computational mathematics where unknown or complicated functions are hoped to be representable by simple ones. Interpolation is a simple and convenient way of approximation. Polynomials are among the most useful and well known as well as the simplest class of functions.

Aims of the project

The purpose of this investigation is to approximate general functions by the polynomials. We consider the interpolation technique and the least squares fitting. We shall consider the one dimensional case first (1D) and then the multiple dimensions (mainly 2D). The two cases may be considered as two separate projects.

Mathematical ideas used

Polynomials are used to represent discrete data or approximate other functions. We consider the methods and criteria of constructing such polynomials. Both the interpolation and least squares processes essentially form matrix equations (linear system) for polynomial coefficients. We address two methods of comparing such functions.

MATLAB techniques used

This project is about functions and their approximations. The MATLAB commands `interp1`, `interp2`, `polyfit` and `polyval`, and M-files `polyfit2.m` and `polyval2.m` are used. There will be an extensive use of MATLAB's graphics capabilities to visualise different functions, all illustrated by two M-files `intdemo1.m` and `intdemo2.m`.

17.1 1D: Introduction

In one space variable, a degree m polynomial may be written as

$$P_m(x) = C(1)x^m + C(2)x^{m-1} + \cdots + C(m)x + C(m+1),$$

where $\mathbf{C} = [C(1)\ \ C(2)\ \ \cdots\ \ C(m)\ \ C(m+1)]$ is the coefficient vector. In MATLAB, such a polynomial is evaluated by the command `polyval`:

Example 1

$$P_4(x) = 2x^4 - 5x^3 + 3x^2 + 8x - 1,$$

that is, $m = 4$, $\mathbf{C} = [\ 2\ \ -5\ \ 3\ \ 8\ \ -1\ ]$. We can evaluate P_4 at points $\mathbf{x} = -9.0 : 4.5 : 10$ and have a dotted as well as a line plot by

```
>> C = [ 2   -5   3   8   -1 ] ;  % Type in the coefficient
>> x = -9.0 : 4.5 : 10.0 ;        % A vector in [-9,10]
>> y = polyval(C, x)     ;        % Evaluate at x
>> plot (x,y,'ow');hold on;       % [0] Scattered points
>> plot (x,y,'-w') ;              % [1] Plot in solid lines
>> xp = -9.0 : 0.01 : 10.0 ;      % A large vector in [-9, 10]
>> yp = polyval(C, xp)   ;        % Evaluate at xp for plots
>> plot ( xp, yp, ':g');          % [2] Dotted lines (better)
>> axis([-9 10 -2000 17500]);     % Fix axis display
>> CQ = polyfit(x,y,2) ;          % Quadratic fitting at (x,y)
>> y2 = polyval(CQ, xp) ;         % Values of approximation
>> plot(xp, y2, '--r');           % [3] Dashed line plot
>> title('Example 1 -- plotting curves');  hold off
```

Here the solid line plot† is not representative of the polynomial because it is not a curve at all and the polynomial is far from being a straight line. We have to use more points to show it as a proper curve (the dotted line). This is the first important point on plots to be noted in doing this project; see Figure 17.1.

Notation and Convention All approximations are based on a knowledge of given *data points*, but *test points* are artificially chosen for analysis. To avoid possible confusion, we use $\mathbf{x}$, $\mathbf{y}$ (size n) for *data points*, $\mathbf{xi}$, $\mathbf{yi}$ (size T) for *test points* and $\mathbf{xp}$, $\mathbf{yp}$ (size L) for *plot points*. Obviously L should be sufficiently large (say $L \gg n$) but T can be any number. For the Example 1 above, we have used $n = 5$ and $L = 1901$.

---

† Here we assume that the background colour is black. On some systems where this is not the case, type `set(gca,'Color','k')` to reset or replace the 'white' ('ow' and '-w') by 'black' ('ok' and '-k').

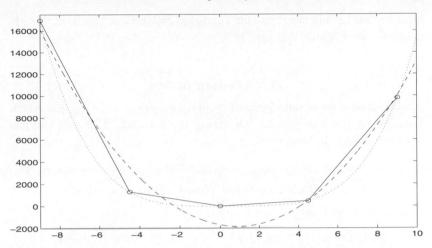

Fig. 17.1. Example 1 – plotting curves (solid lines – polynomial C with few plot points, dotted – with more plot points, dashed – a quadratic fitting).

In practice, we can identify two situations where polynomials are deemed to be useful: (i) given a complicated function, we wish to use a simple polynomial of a certain degree to represent it reasonably; (ii) given a discrete set of points say (x_1, y_1), (x_2, y_2), $\cdots, (x_n, y_n)$, we wish to find a polynomial function to represent the data.

In Chapter 4, we used `polyfit` for fitting a polynomial to case (ii). This method using a single polynomial for approximation is called *a global method*; an example can be seen in Figure 17.1 where a quadratic polynomial approximation is shown. We shall compare such global methods with the so-called *piecewise approximation methods* where several piecewise interpolating polynomials are put together. Here the former uses `polyfit` and the latter uses `interp1`.

17.2 The 1D example M-file `intdemo1.m`

To illustrate `intdemo1.m`, we take $y = \cos(2x)$ in $[-2, 4]$ with $n = 13$
Example 2

$$\begin{cases} \mathbf{x} = -2 \; : \; 0.5 \; : \; 4 \\ \mathbf{y} = \cos(2 * \mathbf{x}). \end{cases}$$

All approximations to this function are compared at 16 points $\mathbf{xi} = -2$: 0.4 : 4 but plotted at 151 points $\mathbf{xp} = -2$: 0.04 : 4. We suggest you print out and study the demo file `intdemo1.m` which is based on this 1D example; see Figures 17.2 and 17.3.

17.3 1D data fitting

Assume that a set of data points (x_1, y_1), (x_2, y_2), $\ldots$, (x_n, y_n) is given. We hope to find a polynomial for fitting the data set. This problem has been considered in Chapter 5.

17.3.1 Global methods (least squares method)

If the polynomial under consideration is of degree m, i.e.

$$P_m(x) = C(1)x^m + C(2)x^{m-1} + \cdots + C(m)x + C(m+1),$$

then the coefficients (vector $\mathbf{C}$) are not known and will be determined by the way we minimize $P_m(x_i) - y_i$. The least squares method proposes to minimise the squares error with regard to $C(i)$s

$$E = E\left(\mathbf{C}\right) = \sum_{i=1}^{n} (y_i - Y_i)^2,$$

where $Y_i = P_m(x_i)$. When $m = 1$, we have the linear approximation $P_1(x) = C(1)x + C(2)$ as discussed in Chapter 5.

The MATLAB commands implementing such a method, for Example 2 with a fifth degree polynomial ($m = 5$), are simply

```
>> x = -2 : 0.5 : 4   ;    % Set up  given vector x
>> y = cos(2*x)       ;    % Set up  given vector y
>> C = polyfit(x,y,5);     % Find out C_i's in P_5(x)
>> xp = -2 : 0.04 : 4;     % Points set up for plot
>> fp = polyval(C, xp) ;   % Evaluate P_m for plot
>> plot(xp, fp)            % Plot the approximating curve
```

17.3.2 Piecewise approximations

You must have realised from Example 1 (and Exercise 17.1) that polynomials or rather global polynomials can be oscillatory or unsatisfactory. We look for alternative approximation methods.

Interpolation is a similar but different approach to least squares fitting. Here we want a polynomial to pass all n data points, i.e. to commit no errors. But the degree m of such polynomials has to be high, namely

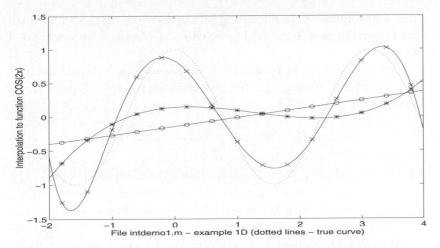

Fig. 17.2. 1D global approximation — linear(o), cubic(∗), order 6(x).

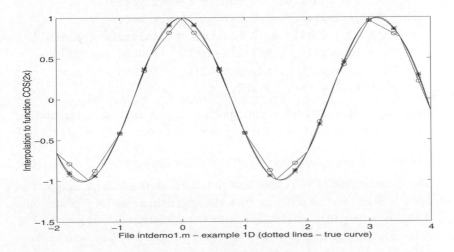

Fig. 17.3. 1D piecewise approximation — linear(o), cubic(∗), spline(x).

$m = n - 1$ (i.e. equal to one less than the number of data points). As global and high order polynomials can be unsatisfactory, interpolation appears to be no better than the least squares method except that the former gives the advantage of exact data representation at data points.

It turns out that polynomials can give good approximations locally (i.e. nonglobally and over a short range), and different or even same order polynomials can be effective if we can divide the data set into small and nonoverlapping sets. This leads to the piecewise approximation approach.†

Therefore, instead of constructing degree $m = n - 1$ polynomials‡ as in Chapter 5, we consider the piecewise approach in the context of interpolation.

17.3.2.1 Piecewise linear and cubic interpolations

The piecewise approach is to use several low order polynomials of interpolation and piece them together (hence the name 'piecewise'). The overall approximation is in general not differentiable,§ though continuous, at the joints, but it is efficient and accurate. Piecewise linear and cubic interpolations are two examples.

The MATLAB command `interp1` can produce both piecewise linear and cubic interpolations, as used below for Example 2:

```
>> x = -2 : 0.5 : 4;    % Set up  given vector x
>> y = cos(2*x);        % Set up  given vector y
>> xx = -2 : 0.04 : 4;  % For interpolations and plots
>> FL = interp1(x,y, xx,'linear');    % Used for plot
>> FC = interp1(x,y, xx,'cubic');     % Used for plot
>> plot(xx,FL,'-r',   xx,FC,'-g')
```

Here **xx** is chosen with a small spacing to plot the two interpolation curves but it can be of any size, e.g. $\mathbf{xx} = -1 : 0.3 : 1$.

17.3.2.2 Continuously differentiable cubic interpolations — splines

The shortcoming of piecewise interpolations or the lack of differentiability is overcome by insisting that the approximation have continuous derivatives (say first and second order) at all interpolation points. For cubic approximations, this gives rise to the popular *spline interpolation* widely used in solving modern engineering problems.

The MATLAB command `interp1` can generate a cubic spline interpolation for Example 2 by

```
>> x = -2 : 0.5 : 4;    % Set up  given vector x
>> y = cos(2*x);        % Set up  given vector y
```

---

† The modern finite element method is one of the best application examples of a piecewise approximation approach.

‡ Interested readers can consult any numerical analysis textbook for details and formulae. Test this out using `polyfit`!

§ If one insists, the overall approximation can be made differentiable; see §17.3.2.2.

```
>> xp = -2 : 0.04 : 4; % Points set up for plot
>> FS = interp1(x,y, xp,'spline'); % Values for plot
>> plot(xp,FS,'-b')
```

17.4 How accurate is my approximation?

If an approximation, when plotted against the exact function,† almost overlaps it, then this is a sign of a good fit. Here we discuss an analytical way of measuring approximation. We shall use and compare it at some test points.

Suppose that $\tilde{x}_1, \tilde{x}_2, \ldots, \tilde{x}_T$ (in vector **xi**) are our test points with the known values $\tilde{y}_1, \tilde{y}_2, \ldots, \tilde{y}_T$ (in vector **yi**). Then if f_is (in vector **fi**) are the values of our approximation at **xi**, we may check how the method performs by calculating (the so-called root mean square error)

$$Error = \sqrt{\frac{1}{T}\|\mathbf{fi} - \mathbf{yi}\|_2^2} = \sqrt{\frac{1}{T}\sum_{i=1}^{T}(f_i - \tilde{y}_i)^2}.$$

If this error is small enough, we say the underlying approximation is good. In theory, we could have $\tilde{x}_i = x_i$ for all i (i.e. test points and data points coincide with $T = n$), but it may not make much sense as with interpolation $Error = 0$. (Think about the definition of 'interpolation'.)

For Example 2, with a least squares approximation by a fifth degree polynomial, we may proceed as follows to estimate the approximation error in root mean square norm.

```
>> x = -2 : 0.5 : 4;     % Set up given data point x
>> y = cos(2*x)      ;   % Set up given data point y
>> C = polyfit(x,y,5);   % Find out C_i's in P_5(x)
>> xi = -2 : 0.4 : 4;    % Set up test points
>> fi = polyval(C, xi);  % Approximations at test points
>> yi = cos(2*xi);       % Exact values at test points
>> T = length(xi);       % Get the size of "xi"
>> Error = sqrt( norm(fi - yi)^2 / T)   % Root mean square
```

Here recall that `norm` is a MATLAB command. We remark that in the last line, if `fi` is replaced by a vector generated from `interp1` (e.g. **FL**),

---

† In real life where the exact function or solution is not available, one usually compares with a known and accurate approximation (which may be expensive to obtain in terms of time).

we have to check if we need to transpose it before taking the difference†
fi − yi.

17.5 Introduction to multi-variable approximation

In §17.1 with 1D, we considered various ways to find a polynomial $P_m(x)$
(defined by its coefficient vector **C**). However, real world problems often
involve more than one space variable; a general m^{th} degree polynomial
in n variables ($\mathbf{R}^n$) may be represented by

$$P_m = P_m(x_1, x_2, \cdots, x_n) = \sum_{i_1+i_2+\cdots+i_n \leq m} C_{i_1 i_2 \cdots i_n} x_1^{i_1} x_2^{i_2} \cdots x_n^{i_n}$$

$$= C_{m0\cdots0} x_1^m + \cdots + C_{00\cdots0},$$

where $i_k = 0, 1, \ldots, m$ for $k = 1, 2, \ldots, n$. For simplicity, we shall concentrate on 2D problems with $n = 2$ and on two special polynomials
with $m = 1$ (linear) and 3 (cubic).

In the case of $m = 1$, we consider the so-called bilinear function of the
form

$$Z(x, y) = a + bx + cy + dxy,$$

while in case of $m = 3$, we approximate functions by the so-called bicubic
function

$$Z(x,y) = a + bx + cy + dxy + d_1xy^2 + d_2x^2y + e_1x^2 + e_2x^3 + f_1y^2 + f_2y^3.$$

In practice, similarly to the 1D case, we can identify two situations
where polynomials are deemed to be useful: (i) given a complicated
function $z = z(x, y)$, we wish to use a simple polynomial $Z = Z(x, y)$
of certain degree ($m = 1$ or 3) to represent it reasonably (based on N
points); (ii) given a set of discrete data points (x_1, y_1, z_1), (x_2, y_2, z_2),
$\ldots$, (x_N, y_N, z_N), we wish to find a polynomial function $Z = Z(x, y)$
to represent the data.

We shall first investigate ways of finding such a polynomial and then
study methods of measuring how good each fit is.

You will need two main M-files specially developed for the remaining
part of this project:

```
polyfit2.m - Least squares fit in 2D (Linear and Cubic)
polyval2.m - Evaluation of "polyfit2.m" (Linear & Cubic)
```

† More precisely, with MATLAB V.4, FS = interp1(x,y, xp, 'spline') always
  produces a column vector even when **xp** is a row vector! This is not so with
  MATLAB V.5.

together with example M-files `intdemo2.m`, `cont4.m` and `cont7.m`.

Notation and convention The careful reader may have noticed that
- m the degree of polynomials;
- n the number of independent variables;
- N the number of data points.

17.6 The 2D M-file `intdemo2.m`

To illustrate the above M-files, a demonstration M-file has been prepared for your reference, where the following example is used ($N = 25$):

Example 3

$$\begin{cases} x = 0 \; : \; 0.25 \; : \; 1 \\ y = 0 \; : \; 0.25 \; : \; 1 \\ z = r \ln(r), \end{cases}$$

where $r = \sqrt{(x-1/2)^2 + (y-1/4)^2}$. As before, you should run and print out this M-file `intdemo2.m` before proceeding.

17.7 Contour plots, 3D plots and slicing

As we are to work with functions of $n - 2$ variables, we should investigate how to visualise such functions.

In MATLAB, we can draw contour plots and 3D plots, both involving similar preparations. Suppose that we want to draw the function

$$z = z(x, y) = (x^6 - y^3 - 0.5) \exp(-x^2 - y^2)$$

in $\Omega = [-2, \, 2] \times [-2, \, 2]$. The command `meshgrid` is first used to set up a grid, followed by calculations of values (if they are not yet known):

```
>>  x0 = -2 : 0.4 : 2;
>>  y0 = -2 : 0.25 : 2;          %% x0, y0: 2 1D vectors
>>  [x y] = meshgrid (x0,y0);    %% To get two matrices
>>  ZZ=(x.^6-y.^3-0.5) .* exp(-x.^2-y.^2); % ZZ = matrix
```

3D plots Two commands `mesh` and `surf` (and their variants†) can be used to visualise **ZZ**; (see Figures 17.4 and 17.5 for actual plots).

```
>>  mesh(ZZ);  title('A mesh plot');
>>  disp('Pause ...'); pause
>>  surf(ZZ);  title('A surf plot')
```

† Type `help mesh` to see how many other commands are referred to. Some commands such as `surf` and `plot3` are only available from MATLAB V.4.

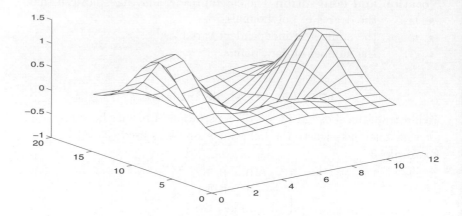

Fig. 17.4. A mesh plot.

To rotate a graph, try the command `view`, e.g. `view(20, 70)`.

Contour plot The command `contour` may be used to draw contour plots but we need to specify values where contour† lines are requested (refer also to the M-file `cont4.m`)

```
>> M=max(ZZ);   mi=min(ZZ);          % Row vectors
>> V=[mi mi/2 0 M/2  M];             % Some heights of ZZ
>> C1 = contour(x,y,ZZ, V,'-r');     % MAIN step
>> clabel(C1);                       % Labels (numbers)
>> title('Contour plot of z(x,y)')
>> C2 = contour(x,y,ZZ, 7,'-g');     % Simpler 7 levels
```

Figure 17.6 shows the result of the first four lines of commands (i.e. use the height vector `V`).

Note: You must have noticed that, with $x0 = -2 : 0.4 : 2$ and $y0 = -2 : 0.25 : 2$, both surface and contour plots appear very nonsmooth. To increase the smoothness, i.e. to view the true curve surfaces, we need to reduce the step-lengths, e.g. use $x0 = -2 : 0.01 : 2$ and $y0 = -2 : 0.02 : 2$ to generate more grid points.

---

† We remark that in the older versions of MATLAB (say Versions 4-), variations of the input method shown here are also acceptable, e.g. `C1 = contour(ZZ,7, x0,y0, '-r')` or `C1 = contour(ZZ,V, x0,y0, '-r')` or `C1 = contour(x0,y0,ZZ,7, '-r')`. Only the very last variation is valid with MATLAB V.5.

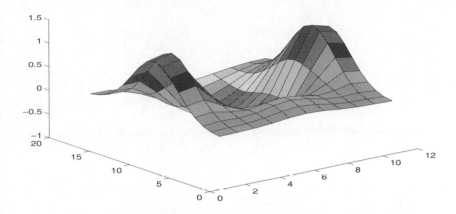

Fig. 17.5. A surf plot.

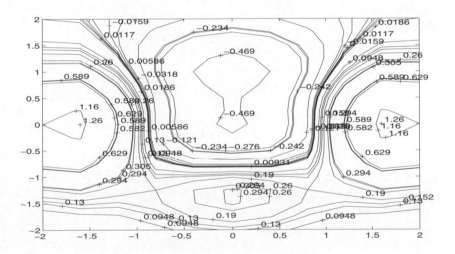

Fig. 17.6. A contour plot.

Sliced plots Sliced plots † are line plots that are sliced off 3D plots in one axis direction. For Example 3, we may slice at mid-intervals in each direction and then plot each slice separately (see Figure 17.7):

---

† MATLAB V.5 introduces a new and convenient command called `slice` for the same purpose. Type `help slice` to find out more details.

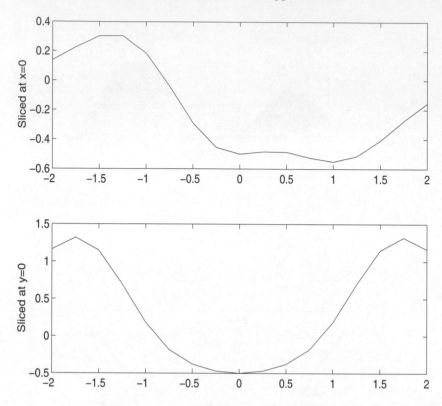

Fig. 17.7. Sliced plots.

```
>> x = 0;    y = -2 : 0.25 : 2;    %% x=value, y=vector
>> Zx=(x.^6-y.^3-0.5) .* exp(-x.^2-y.^2);   % Zx=vector
>> subplot(211);plot(y, Zx,'-r'); title('Sliced plots')
>> ylabel('Sliced at x=0')
>> y = 0;    x = -2 : 0.25 : 2;    % Fix y
>> Zy=(x.^6-y.^3-0.5) .* exp(-x.^2-y.^2);    % Zy=vector
>> subplot(212); plot(x, Zy, '-g');
>> ylabel('Sliced at y=0')
```

A different (and maybe better) method of plotting sliced plots is by use of the command plot3; we repeat the above example as follows:

```
>> subplot(111)
>> x = 0;    y = -2 : 0.25 : 2;
>> Zx=(x.^6-y.^3-0.5) .* exp(-x.^2-y.^2) ;
```

```
>> h1=plot3(zeros(size(y)), y, Zx, '-r');
>> hold on                    %%% Necessary
>> y = 0;     x = -2 : 0.25 : 2;
>> Zy=(x.^6-y.^3-0.5) .* exp(-x.^2-y.^2) ;
>> h2=plot3(x, zeros(size(x)), Zy, '-g');
>> zlabel('Slices at planes x=0 and y=0')
>> grid on; hold off % Change viewpoint below
>> view(10,50)        % Try view(80,50) or view(175,50)
>> set(h1,'Linewidth', 3); set(h2,'Linewidth', 2)
>> set(gca,'Ytick', -2:1:2)
```

Here the last two lines annotate the graph in an advanced way; type
get(gca) and help set to find out which other attributes of a graph can
be modified. As sliced plots do not represent the full graph, they should
be used with caution but they can be useful when comparing several
approximations. Of course, slicing can be in any suitable directions, e.g.
along $x + y = 1$.

17.8 The '\' global method

Suppose $Z(x, y)$ is our approximation to the true function $z(x, y)$ and
$z_i = z(x_i, y_i)$; it may be possible that $z(x, y)$ is not known but that its
discrete values are known.

With the least squares approach, we hope to minimise the squares
error (assume $Z_i = Z(x_i, y_i)$ for all i)

$$E_2 = \sum_{i=1}^{n} (z_i - Z_i)^2.$$

The minimisation is with respect to the underlying parameters; e.g.
a, b, c, d for the bilinear case of §17.5. The method has been implemented in two M-files where the '\' approach has been used (see the
appendix to Chapter 5 for a theory). The syntax to use polyfit2.m
and polyval2.m for Example 3 is as follows (see cont7.m):

```
x0 = 0:0.25:1;  y0 = x0;      %% Interpolation nodes
[x y] = meshgrid(x0, y0);
 R = sqrt ( (x-1/2).^2 + (y-1/4).^2 );
 z = R.*log(R+eps) ;          %% True and given values
 C1 = polyfit2( x,y,z, 'linear');
 C2 = polyfit2( x,y,z, 'cubic' );
%%%%%%%%%%%%%%%%%%%%%%%% Evaluations for plots only %%%%%%
```

```
    x0 = 0:0.02:1;              y0 = x0 ;           % set nodes
    [xi yi] = meshgrid(x0, y0);
    Z1 = polyval2( C1, xi,yi, 'linear');     % Bilinear
    Z2 = polyval2( C2, xi,yi, 'cubic' );     % Bicubic
%%%%%%%%%%%%%%%%%%%%%% Contour Plots %%%%%%%%%%%%%%%%%%%%%%%
    contour(Z1,9, x0,y0, '-g');     hold on; % 9  levels
    contour(Z2,9, x0,y0, ':r');     hold off
```

17.9 The piecewise method

The global method tends to fail for certain difficult functions because high order polynomials are oscillatory. One alternative, as in 1D, is to use piecewise methods that have the advantage of smoothness of low order polynomials and reasonable local accuracy.

Here we shall consider the interpolation method using piecewise low degree polynomials. The MATLAB command `interp2` can be used for both bilinear and bicubic interpolations. As we know, the intersection of two pieces of surfaces is a curve that has an uncountable number of points. It is not possible to have continuity and differentiability† over this curve, because the number of freedoms of any degree polynomial is limited. Hence there are no 'splines' in other than one dimension.

We now show how to use `interp2` for Example 3 (see `cont7.m`)

```
>> x=0 :0.25: 1;     y=x;                [x y]=meshgrid(x, y);
>> R=sqrt ( (x-1/2).^2 + (y-1/4).^2 ); z = R.*log(R+eps);
>> xa=0 :0.05: 1 ;     ya=xa ;           % New 1D vectors
>> [xi yi] = meshgrid(xa, ya) ;          % New 2D matrices
>> zL = interp2(x,y,z, xi,yi, 'linear');
>> zC = interp2(x,y,z, xi,yi, 'cubic');
>> contour(zL,9, xa,ya, '-w'); hold on;
>> contour(zC,9, xa,ya, '-b'); hold off
```

17.10 Comparison of approximations

For a function $z = z(x, y)$, we have discussed four kinds of approximations $Z = Z(x, y)$. We hope to decide which one is the closest to the original function by comparing them at some selected points.

---

† However, in such cases, one may be content with incomplete differentiability, i.e. differentiability at some points (not all points of a curve). Tensor products can achieve such a result.

These selected points, called test points as in 1D, can be those used to generate various plots but should not be the starting data points x0 and y0.

Suppose that an approximation Z and its corresponding true function values z are in the form of matrices (say of size $N_1 \times N_2$). In the example of the last two sections, such an approximation Z may be $Z1, Z2, zL, zC$ and z is calculated using the original function $z(x, y)$. Then the following two methods can be considered.

Geometric comparison Using the `hold on` option, we may overlap an approximation Z on the contour plot of the true values z (using different colours of course). Then large discrepancies would indicate that the underlying approximation is not very good and vice versa. *Note* it is not easy to overlap plots with `mesh` and `surf`.

One may also generate a contour plot and decide where the main features of a function are, before producing suitable sliced plots for comparisons. Conclusions drawn from a geometric comparison can be subjective but it is effective to identify the poorest approximation method.

Algebraic comparison A very useful but boring method for measuring discrepancies between two matrices is by use of norms, e.g. $E = $ `norm`$(Z - z)$ as an error indicator† — a small E indicates good approximation and vice versa.

In engineering applications, as in 1D, it is more common to use the so-called root mean squares norm defined by

$$E = \sqrt{\frac{1}{N_1 N_2} \sum_{i=1}^{N_1} \sum_{j=1}^{N_2} (Z_{ij} - z_{ij})^2}.$$

The smaller such an error is, the better the method is.

Luckily the above formula resembles the Frobenius norm of matrix $(Z - z)$ that can be computed via the MATLAB command `norm(Z-z, 'fro')`. Thus the quantity E can be computed as follows

```
>> [m  n] = size(Z);    % Get the dimension of matrix
>> mn = m * n ;         % Total number of nodes
>> format compact;      % Don't leave extra space lines
>> disp('Root Mean Squares error is :');
>> E - sqrt( norm( Z-z, 'fro')^2 / mn )
```

Finally, we remark that data fitting and function approximations play

† See Chapter 16 for more examples of `norm`.

important roles in scientific computing. As this project shows, an efficient method may not work well for all problems — a fact that typifies the difficulty and challenge for mathematicians doing real modelling and simulations. There are many other choices of global functions that may be substitutes for polynomials such as trigonometric functions, radial basis functions and, more recently, wavelet functions. Again for each case, one may look at some kind of piecewise versions — functions with relatively small compact support. Overall, the design of a robust method involves careful experiments and observations, and above all, mathematical analysis.

Exercises

17.1   Given the function $f(x) = 1/(1+25x^2)$ in $[-1, 1]$ and nine data points $\mathbf{x} = -1 : 0.25 : 1$, plot the function against its polynomial approximations of degrees 3 and 7.

17.2   Following Exercise 17.1, plot the original function against its piecewise linear, cubic and spline approximations.

17.3   For the following given functions

   (a) $f(x) = -(x+a)^R + [a^R(1-x) + (1+a)^R x]$ with $a = 10^{-2}$ and $R = -\frac{1}{4}$ in $[0, 1]$,

   (b) $f(x) = (1-x)\{\tan^{-1}[b(x-x_0)] + \tan^{-1}[bx_0]\}$ with $b = 100$ and $x_0 = 1/4$ in $[0, 1]$,

   (c) $f(x) = \tanh[20(x - 1/2)]$ in $[0, 1]$,

   (d) **EXTRAS** $f(x) = 10000(f_1(x) + f_2(x))|(x-1/4)(x-3/4)|^3$, with $f_1(x) = -(x+a)^R + [a^R(1-x) + (1+a)^R x]$ and $f_2(x) = -(1-x+a)^R + [a^R x + (1+a)^R(1-x)]$ where $a = 0.3$ and $R = -1/4$, in $[0, 1]$,

   (i) Compare the performance of four approximations: least squares method using `for m = 1 : 20`, piecewise linear method, piecewise cubic method and piecewise cubic spline method by using $n = 10$ uniformly distributed data points ($\mathbf{x} = 0 : 1/9 : 1$);

   (ii) Try to improve the performance of at least one of the two approximations: least squares method using `for m = 1:20` and piecewise linear method, by selecting $n = 10$ manually placed data points that are nonuniform (say $\mathbf{x} = [0 : 0.1 : 0.4 \quad 0.6 : 0.1 : 1]$ or $\mathbf{x} = [0 \quad 0.26 : 0.15 : 0.44 \quad 0.47 :$

0.035 : 0.65 1]). Plot the two approximations on the same graph.

Notes In each case, to simplify the problem, you should

(1) test the root mean squares error at $T = 10n+1$ uniformly distributed test points (i.e. about 1/10 spacing of data points);

(2) rank the methods from the best to the worst in a performance table (according to error indicators at test points);

(3) plot all competitors on the same graph.

Hints:

(I) For the least squares method, you should try up to the degree of $m = 20$ polynomial but only quote the best result for comparison.

(II) For the case of using 'manually placed data points', if you are not sure about how to find these points, the following may help

• plot a given function,
• plot the first derivative of a given function,

then place more points near large variations of the function or near large values of its derivative.

(III) $\dfrac{d\tanh(x)}{dx} = 1 - \tanh^2(x)$ and $\dfrac{d\tan^{-1}(x)}{dx} = \dfrac{1}{1 + x^2}$.

(IV) On completion, you should obtain

(1) 6 graphs;

(2) 3×4 sets of root mean squares errors and 3 orderings from using uniform points;

(3) 3×2 sets of root mean squares errors and 3 orderings from using nonuniform points.

17.4 Plot the following two functions and also show their sliced plots at some suitable positions (only one 3D and one slice plot are required for each case):

• $g_1(x, y) = \tanh[20(x + y - 1)]$ in $[0, 1] \times [0, 1]$;
• $g_2(x, y) = \tanh[20(x^2 + y^2 - 1/2)]$ in $[0, 1] \times [0, 1]$.

17.5 For the two functions in Exercise 17.4, show contour plots (with five levels) of their linear and cubic approximations; use $N = 11$ points in each coordinate direction, i.e. x0 = y0 = 0 : 0.1 : 1.

17.6   For the two functions in Exercise 17.4, show contour plots (with
        five levels) of their piecewise linear and cubic approximations;
        use $N = 9$ points in each coordinate direction, i.e. x0 = y0 =
        $0 : 1/8 : 1$.

17.7   For one of the two functions in Exercise 17.4, compare its global
        linear and *cubic* approximations based on $N_1 = N_2 = 13$ points
        (see Exercise 17.5 with x0 = y0 = $0 : 1/12 : 1$) against its
        piecewise *linear* and *cubic* approximations based on $N_1 = N_2 =$
        8 points (see Exercise 17.6 with x0 = y0 = $0 : 1/7 : 1$):

        (a) First carry out algebraic comparisons and rank the four
            methods in each case from the best to the worst. The
            number of test points should be more than $4 * N_1 * N_2$.
        (b) Then present suitable graphs supporting your conclu-
            sions. (Definitely sliced plots are effective but try to get
            at least one contour or a 3D plot as well.)

18

Ordinary Differential Equations

Aims of the project

You are invited to study the solutions of a list of ordinary differential equations using whatever methods you have at your disposal.

Mathematical ideas used

Some of the equations are capable of analytic solution using such mathematical tools as: separation of variables, integrating factors or series solutions. Others are examples of homogeneous, or constant coefficient differential equations. What you can bring to bear will very much depend on your mathematical background at this point.

MATLAB techniques used

All the numerical techniques required have been introduced in Chapter 7. For example, grain (or phase) plot analysis and the numerical solution of coupled first order equations. You may find helpful the M-files associated with that work (`fodesol.m`, `species.m`, `vderpol.m`,...). You can use these directly or copy and modify them as required.

18.1 Strategy

Your aim, for each equation in the list of exercises, is to provide the following information as appropriate:

(a) For first order equations, a grain plot with typical solutions superimposed. For second order, or coupled first order equations, sketch of a typical phase plot.

(b) An analytic general solution *if* you can find one.

(c) The particular solution for the specified initial conditions.

(d) Any other comments you wish to make on the nature of the solutions, their stability etc.

In each case, start by classifying the type of the differential equation. Is it linear? What order is it? Has it got constant coefficients? Is it homogeneous? If it is of a type which you recognise, then try to solve it 'analytically', that is, by paper and pencil. If you can't, then see if you can use one of the techniques discussed in Chapter 7. If you have been able to solve the equation exactly, you can compare the solution with the numerical or graphical one got by using MATLAB. In each exercise, you are given a *particular* solution to find. You should try your best to find out as much as you can about all possible types of solution.

Exercises

18.1   Find y at $x = 1$ for the particular solution of

$$\frac{dy}{dx} = -2xy$$

satisfying the initial condition $y(0) = 1$.

18.2   Find y at $x = 1$ for the particular solution of

$$\frac{dy}{dx} = e^{-2xy}$$

satisfying the initial condition $y(0) = 1$.

18.3   Find y at $x = 5$ for the particular solution of

$$(x - 2\sqrt{xy})\frac{dy}{dx} - y = 0, \qquad (x > 0, y > 0)$$

satisfying the initial condition $y(1) = 10$.

18.4   Find y at $x = 1$ for the particular solution of

$$\frac{d^2y}{dx^2} + 3\frac{dy}{dx} + 2y = e^x$$

satisfy the initial condition $y(0) = 1$, $y'(0) = 0$.

18.5   Find y at $x = 4$ for the particular solution of

$$\frac{d^2y}{dx^2} + y^2 - x^2 = 0$$

satisfying the initial condition $y(0) = 1$, $y'(0) = 0$.

18.6   Find y at $x = 4$ for the particular solution of

$$\frac{d^2y}{dx^2} = \cos(2x)y$$

satisfying the initial condition $y(0) = 1$, $y'(0) = 0$.

18.7   Find y at $x = 4$ for the particular solution of

$$x^2\frac{d^2y}{dx^2} + x\frac{dy}{dx} + x^2y = 0$$

satisfying the initial condition $y(0) = 1$, $y'(0) = 0$. (*Hint:* You may have difficulty finding the numerical solution when x is near 0 (called a 'regular singular point'). By considering a series (Maclaurin) expansion such as $y(x) \approx a + bx + cx^2$ you can work out how y and its derivative must behave for very small x-values should your M-file need to know this.)

18.8   Find x at $t = 1$ for the particular solution of

$$\frac{dx}{dt} = x(2x + y - 3), \quad \frac{dy}{dt} = y(x + 2y - 3)$$

satisfying the initial condition $x(0) = \frac{4}{5}$, $y(0) = 1$. (*Hint:* Think also about fixed points and stability as studied in Chapter 7.)

18.9   Find y at $x = 1$ for the particular solution of

$$(1 - x^2)\frac{d^2y}{dx^2} - 2x\frac{dy}{dx} + 12y = 0$$

satisfying the initial condition $y(0) = 0$, $y'(0) = -\frac{3}{2}$. (*Hint:* See the advice given for exercise 18.7, if you experience difficulty in finding solutions at particular x values.)

Part three
Modelling

19

Checkout Queues: Long or Short

In daily life, we meet many examples of first-in-first-out or *fifo* queues that are usually *simple queueing systems*. A more complex *multiqueueing system* might on one hand involve a *fan-in* structure in which several separate queues have to merge at a later facility, on the other hand, involve a *fan-out* structure where a single queue branches to form distinct subqueues. The people, objects, customers that flow through the system, known as *entities*, can have *attributes*, such as requiring leaded or unleaded petrol, paying by cash or cheque, that they carry with them through the system.

Aims of the project
The purpose of modelling queues is two-fold. Firstly, as customers, we tend to prefer short queues in order to save time. Secondly, as business managers, we hope the service utility (i.e. the ratio of actual time when a service is utilised over the maximum available service time) is approaching 100% to maximise profits but still want to avoid long queues for the sake of customer loyalties. Therefore it is of practical interests to predict the peak times of long queues before deciding how to improve the service.

Mathematical ideas used
Statistical distributions and probabilities used here have been discussed in Chapter 6. Sparse vectors of 0s and 1s, for data extraction, are generated by various conditions. Flexible vector operations will be needed.

MATLAB techniques used
The supplied M-file `queue.m` is mainly used, assisted by three other M-files `exprand.m`, `normrand.m`, `unirand.m`. There will be an extensive use of MATLAB commands `plot`, `bar`, `sort`, `rand`.

This project is divided into three parts. §19.1 deals with some tasks of modelling simple queues and introduces the main M-file `queue.m`. §19.2 models queues at a busy motorway petrol filling station while §19.3 considers queues at the Leo's cafe in a retail park. Both problems, each representing a separate project, involve making decisions on staffing levels, and improving profit margins and promotion issues.

19.1 Simulating queues

A simulation is performed as a sequence of *events* which are of two types

- *Arrival:* An entity arrives at a service and joins the queue. If the queue was previously empty and the service idle then the entity service starts immediately, otherwise the entity joins the back of the queue and waits in line.
- *End-of-service:* An entity moves on to next stage. If there are no entities waiting in the queue then the next service starts immediately and is not noted as a separate event, as far as the current entity is concerned. Of course, if the current queue is empty then the service is unoccupied until the next arrival.

If there is a further stage, an entity moves directly to the next queue and this arrival at the new queue is not a separate event. If there are no further stages, the entity leaves the system. In the examples of the next section, 'vehicle' = entity and 'petrol pump' = service while in §19.3, 'people' = entity and 'travel from stores to cafe' = service.

All entities have to be separated into two distinct queues at some stage. In the case of a single queue and a single server, we typically need

- **vector 1** — a sequence of inter-arrival times;
- **vector 2** — a sequence of service times (attributes).

Note that with `cumsum` inter-arrival times are converted into arrival times. As we said one cares both about an entity (or entities) and about the overall utility of a service, so as to make the underlying business convenient and efficient as well as cost effective. The above two vectors can be used to work out the required information — we have developed a M-file `queue.m` for such a purpose. Firstly, recall how we model the inter-arrival times and service times.

19.1.1 The statistical theory

The inter-arrival time I_t may be described by a negative exponential distribution. As in Chapter 6, this can be denoted by

$$I_t = -L\ln(1 - r),$$

where r is the uniform random variable, L is the mean value for inter-arrivals, and I_t is the inter-arrival time (simulated).

Service times are usually based on simple discrete attributes (such as 40% turn left and 60% turn right) and can be simulated by MATLAB's random numbers as in Chapter 6.

19.1.2 The M-file queue.m and some associated M-files

The supplied M-file queue.m is designed to work out the waiting time and end-of-service time, given an arrival time and a service time. The syntax is as follows

```
>>   [Mean_q Serv_u Wait_t Stop_t] = queue(Arr_t, Ser_t)
```

where the two input vectors are

- **Arr_t** the vector of arrival times;
- **Ser_t** the vector of service times;

and the four output parameters are

- Mean_q a scalar showing the mean queue length for the period of the simulation;
- Serv_u a scalar showing the fraction of total time that the service was in use;
- **Wait_t** the vector of waiting times (queueing to be served), excluding the service times;
- **Stop_t** the vector of end-of-service times.

For example, with two arrivals **Arr** $= [2 \ 8]$ and corresponding service time **Ser** $= [7 \ 4]$ (in minutes), we have

```
>> [M  S  Wait  Stop] = queue(Arr, Ser)
   M = 0.9231
   S = 0.8462
   Wait =  0  1 % No waiting for 1st but only for 2nd
   Stop =  9 13   % The time when service is finished
```

Table 19.1. *Simulation data of a supermarket checkout.*

Inter-arrival time **IT**	Number of items **N**	Service time **ST** = $S * $**N**
47	9	90
94	7	70
58	7	70
103	3	30
3	6	60

The M-file `queue.m`, as used above, has been devised to simulate a simple queue. For more complicated examples it is necessary to write a simple program that calls `queue.m` more than once.†

To simplify the generation of random numbers, you may use the following three M-files: `exprand.m`, `normrand.m` and `unirand.m`; these M-files should be easy to understand as illustrated in Chapter 6 but type `help` to see the usage if necessary. To compute a matrix of $n \times k$, exponentially distributed with mean $\mu = mu$, type

```
>> matrix = exprand(mu,n,k)
```

To compute a matrix of size $n \times k$, normally distributed with mean $\mu = mu$ and standard deviation $\sigma = sigma$, type

```
>> matrix = normrand(mu,sigma,n,k)
```

To compute a matrix of size $n \times k$, uniformly distributed between a and b, type

```
>> matrix = unirand(a,b, n,k)
```

19.1.3 The best time to go shopping

Customers arrive at a supermarket express checkout according to a negative exponential distribution of inter-arrival times with mean $L = 50$ seconds. They have $I = 10$ items or less and the checkout time is $S = 10$ seconds per item. Here the inter-arrival time of customers can be simulated by **IT** = `exprand(5, L)` for five arrivals, and the number of their purchases is simulated by N = $\mathtt{ceil(unirand}(5, 0, 10))$. Some sample data are shown in Table 19.1.

---

† We remark that from an efficiency point of view, service utility should reach 1.0 (or 100%) to minimise cost, but this inevitably means long queues that should be avoided to maximise customer satisfaction and loyalty.

The arrival time **AT** can be worked out using `cumsum` based on the inter-arrival time **IT**. Further we can work out when a person leaves the supermarket till, how long each person needs to wait in the queue, the average waiting time and the percentage of time when the till is in full use. All this information can be obtained by a single call to the M-file `queue.m` as follows

```
>> IT = [47  94  58  103  3];
>> AT = cumsum( IT )
>> ST = [90  70  70  30  60];  % Service time
>> [Mean_q  Ser_u  Wait_t  Stop_t] = queue(AT,ST)
```

The solution from MATLAB will be

```
      AT = 47    141   199   302   305
   Mean_q = 0.9158
    Ser_u = 0.8163
   Wait_t =  0     0    12     0    27
   Stop_t = 137   211   281   332   392
```

This means that on average a customer finds a queue of length `Mean_q` = $0.9158 \approx 1$ person waiting for service (i.e. the till service is fast). In particular of the five arrivals, the first, second and fourth person need not to wait but the third and the fifth person are a bit unfortunate having to wait in the queue for one person ahead.

19.1.4 An event table

Once a simulation is performed, we can construct an *event table* based on the information obtained. An event table can show the full details of each entity: arrival time, details of attributes, queue length and end-of-service time. Thus we may keep a track record of events to facilitate the efficient running and management of a service.

For the above shopping example, most queueing observations can be reflected in an event table (see Table 19.2) which is built up from arrival time, waiting time, and stopping time etc, where N denotes the number of items and EoS the end of service time while QL represents the queue length that is calculated by hand (*note:* the mean of these QLs has been given by `queue.m` and the last column EoS is equal to the stopping time as given by `queue.m`).

Table 19.2. *Simulation results of a supermarket checkout.*

Event	Clock (seconds)	N	ST	QL	Next arrival	EoS
start	0			0	47	free
arrival(1)	47	9	90	1	141	137
arrival(2)	141	7	70	1	199	211
arrival(3)	199	7	70	2	302	281
arrival(4)	302	3	30	1	305	332
arrival(5)	305	6	60	2	343	392

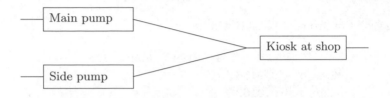

Fig. 19.1. Layout of the filling station (Main = motorway).

19.2 The motorway filling station

The motorway filling station (see Figure 19.1) stands at a road junction with one entrance from the motorway and one from a side road. There is one petrol pump for each entrance with cars from the motorway queueing at one pump and cars from the side road queueing at the other.

All customers queue to pay at a single cash point after serving themselves with petrol, where 20% pay by cash, 30% by credit card and 50% by cheque (think how you could distinguish these). The inter-arrival times of cars at each entrance are distributed according to a negative exponential distribution, the mean inter-arrival times being 100 seconds on the motorway and 150 seconds on the side road. The service times at each pump are distributed uniformly, ranging from 10 seconds to 100 seconds at the motorway pump and from 30 seconds to 150 seconds at the side road pump. The time taken at the cash kiosk is 30 seconds for payment by cash, 40 seconds for payment by credit card and 60 seconds for payment by cheque.

Based on a small sample of the first few arrivals, we can determine their arrival times, service times and cash point times (to the nearest second) by using pseudorandom numbers as shown in Table 19.3, where main refers to the motorway pump and the random vector **r** is used to

Table 19.3. *Sample data for the motorway filling station.*

| Arrival **AT** | | Service **ST** | | Payment method | | Time taken |
main	side	main	side	random r	symbol s	at cash point
2	17	56	46	0.96	+1	60
246	72	52	129	0.96	+1	60
344	98	94	74	0.07	−1	30
400	356	33	19	0.22	0	40
	405			0.11	−1	30

work out the payment method (vector **s**). All times, shown in tables of this section, are converted into seconds. Further in Table 19.4, we can construct the corresponding event table to display the results (*note:* EoS = end-of-service corresponds to the event in column 1). Again we may use queue.m to verify the correctness of this EoS column as follows:

```
>> Arr = [ 58 63 201 275 298]; %% Arrival at kiosk
>> Ser = [ 60 60  30  40  30];
>> [mq  su  wait_t  EoS]  =  queue(Arr, Ser)
```

where **Arr** and **Ser** are collected from the '*.k' rows (i.e. the actual combined arrivals at the cash kiosk) of Table 19.4. The results (for EoS) match those corresponding entries in the last column of Table 19.4, as expected:

```
[mq su] = 0.8464  0.6377
  wait_t =   0    55    0    0   17
     EoS = 118   178  231  315  345 % Good %
```

Note: For the cash point times, if r denotes a vector of uniformly distributed numbers in $[0, 1]$, let customers pay by cash if $0 \leq r \leq 0.2$, by credit card if $0.2 < r \leq 0.5$ and by cheque if $0.5 < r \leq 1$. A simple method of generating 1s for those $r < 0.2$ and 0s for those $r \geq 0.2$ is the following $k1 = (r < 0.2)$.

19.3 The Leo's cafeteria

The Leo's cafeteria is situated on the exit route in a busy retail park, in a city centre, where there are three main businesses: a superstore, a DIY store and a garden centre. Customers visit these businesses first before dropping into the cafe for some food and drink. As illustrated in Figure 19.2, customers arrive at one of three businesses to buy various

Table 19.4. *A sample event table for the motorway example ('m1.p' means the first arrival 'm1' at the motorway pump (referred to as main), 'm1.k' means the arrival of 'm1' at the kiosk, and similarly 's1.p' and 's1.k' are for the side road).*

Event code	Clock (s)	Pay time	Queue length (QL)			Next arrival		EoS (exit)
			main	side	kiosk	main	side	
start	0		0	0	0	2	17	
m1.p	2		1	0	0	246	17	58
s1.p	17		1	1	0	246	72	63
m1.k	58	60	0	1	1	246	72	118
s1.k	63	60	0	0	2	246	72	178
s2.p	72		0	1	2	246	98	201
s3.p	98		0	2	2	246	356	275
s2.k	201	30	0	1	1	246	356	231
m2.p	246		1	1	0	344	356	298
s3.k	275	40	1	0	1	344	356	315
m2.k	298	30	0	0	2	344	356	345
m3.p	344		1	0	1	400	405	438
s4.p	356		1	1	0	400	405	375

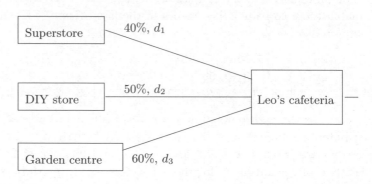

Fig. 19.2. Location of the Leo's cafe.

items and after some time most of customers arrive at the last stop — the cafe (via the front entrance). The cafe has two entrance doors: a front (main) one and a rear one, but only one cash kiosk. Previous data show that:

(a) The inter-arrival times of customers at superstore, DIY store and garden centre all satisfy the negative exponential distribution, with means of m_1, m_2 and m_3 seconds respectively.

    Customers arriving at the front door of the cafe are from one of these three businesses but a smaller number of customers arriving at the rear door of the cafe are mostly from a nearby city centre car park; the inter-arrival time for these people alone (at the rear door) satisfies the negative exponential distribution, with a large mean of m_4 seconds.

(b) 40% of customers at the superstore will visit the cafe next.
50% of customers at DIY store will visit the cafe next.
60% of customers at garden centre will visit the cafe next.

(c) It takes customers from the superstore, DIY store and garden centre (on average) d_1, d_2 and d_3 seconds respectively from the moment they first enter a store to the moment they arrive the cafe.

(d) Of all customers arriving at the cafe, the time taken at the cash kiosk is 20 seconds for payment by cash but 40 seconds for payment by credit card. Actually 90% of people pay by cash and so only 10% by credit card.

Exercises

19.1   Assume the mean value L for inter-arrivals of fishing boats at a sea port is 3 minutes. Consider three kinds of distributions for the inter-arrival time I_t:

    (a) uniform distribution in $[0, 6]$;
    (b) normal distribution with standard deviation $\sigma = 1$;
    (c) negative exponential distribution.

Take the first 100 arrivals in each distribution and, using queue.m with the service time of 2 minutes for each boat, plot all three arrivals versus respective waiting times on the same graph. Find the peak waiting time for each distribution and mark it on the graph.

19.2   For the supermarket example in §19.1.4, taking $L = 45$, $I = 12$, $S = 8$, use queue.m to simulate nine arrivals and complete a similar event table.

19.3   For the motorway filling station of §19.2:

(a) Generate 145 arrivals on the main road (motorway) and 96 arrivals on the side road, and perform a simulation before working out all three waiting times (at two pumps plus cash kiosk) with the help of `queue.m`. Further:

  - obtain bar charts (or other forms of suitable plots) to present the information on arrival times against the waiting times (for each of the pumps and for the cash kiosk);

  - work out a peak time (in terms of the longest waiting time) for each pump and also for the cash kiosk. Mark the peak time on the respective plots.

Hints The results will be in three vectors of waiting time. The event table is not required here but do state the methods used. The arrival time at the cash kiosk is the combination of the two EoS times (stop-times) from the two pumps. This arrival time must be sorted. In MATLAB, an easy way to combine two row vectors **a** and **b** into **c** is by **c** = [**a b**].

(b) Generate random numbers under each heading and complete (by hand again) an event table similar to Table 19.4 until a clock time of at least 500 seconds.

Hint Do a large simulation and then find out how many are actually needed for a clock time close to and above 500 seconds.

(c) Assume that in the motorway problem, each customer pays between £2 and £18 in a uniform distribution. Calculate the total income of first half-hour. Suppose that a promotion costing £70 would change the mean inter-arrival time to 80 seconds and 130 seconds for the motorway and side road pumps respectively.

  - Decide if the promotion is worthwhile.

  - Run your M-file n times (where $n \geq 10$) and work out the probability $\mathcal{P}$ that your decision above is correct; ($\mathcal{P} \approx k/n$ if the same decision occurs k times in n trial runs).

Hint For this final question, you should use a variable `seed` for generator `rand`. The following may be used

```
>> use_seed = sum(100*clock);
>> rand('seed', use_seed) ;
```

before each run of your M-file.

19.4 For the Leo's cafeteria example of §19.3 using $m_1 = 44$, $m_2 = 55$, $m_3 = 36$, $m_4 = 100$ and $d_1 = 145$, $d_2 = 108$, $d_3 = 69$:

(a) Generate 100 arrivals for each of the three main businesses and 50 arrivals at the rear door of cafe. Then work out how many customers out of each 100 actually arrive at the cafe. Show all four kinds of arrivals (in minutes) at the cafe on the same graph (using different colours and line patterns).

(b) In (a), what is the total number of customers arriving at the cafe. Plot arrival time (in minutes) versus service time (in seconds) at the cafe. Further show arrival time (in minutes) against waiting time (in seconds) on another graph after calling queue.m. Find the maximum waiting time (or the peak of graph) and mark this point by the symbol 'o' on this second graph.

(c) Following (a) and (b), how many customers of each kind of arrival have arrived at the cafe within the first 40 minutes. What is the total number of customers in the first 40 minutes? Suppose that the amounts of money customers spend in the cafe satisfy a normal distribution with a mean of £1.49 and a standard deviation of 30 pence. Calculate the takings in the first 40 minutes. What is the profit margin if the running cost of the cafe is £90?

Hint 1 For the cash point times, as in §19.2, let customers pay by cash if s = (0<=r & r>=0.9), and by credit card if s = (0.9 < r), where r is the normalised random vector.

Hint 2 The MATLAB command nnz counts the number of nonzeros in a vector. Also, the command sort may be used to pick up all nonzeros in a vector as in

```
>>    arr = [ 2  0  1  5  0  0  4  0  9 ]
>> arr_1 = -sort(-arr), number = nnz(arr)
>> arrival_2 =   sort( arr_1(1:number) )
```

giving *number* = 5 and

```
     arr_1 =  9  5  4  2  1  0  0  0  0
 arrival_2 =  1  2  4  5  9
```

20

Fish Farming

Aims of the project

You are given data on the current distribution in size and age of a particular species of fish on some fishing banks. You are invited to help establish a sensible fishing strategy (size of net mesh and frequency of harvest) so that the stocks of fish can be maintained at a viable level. As part of this, you are required to develop a model for the growth rate of fish.

Mathematical ideas used

Leslie matrices (Chapter 2) are used to describe the evolution of the age distribution and differential equations (Chapter 7) to describe the growth of the fish. Least squares fitting (Chapter 5) is a further optional technique.

MATLAB techniques used

Much of the above analysis can be accomplished with the M-files listed below. You may have to modify some of these in the course of your study. In each case typing `help` will give information on the purpose and usage. You will also make use of standard MATLAB commands such as `eig` (eigenvalues), `plot` and `bar` (bar graphs).

`fishy.m`	–	loads fish data
`leslie.m`	–	example of population evolution (see Chapt 2)
`lmfish.m`	–	basic Leslie matrix for fish population
`fodesol.m`	–	graphical and numerical solution of first order ODE
`mparft.m`	–	multi-parameters least squares fit
`mparst4.m`	–	example set up for the above
`resid4.m`	–	residuals function for the above
`fishdat.m`	–	uses fishy.m to load data for above fit

257

20.1 Preliminary look at the problem

Use the command `fishy` to load some data on the fish. Typing `help fishy` will show you what is there. You could use the usual `plot` command to have a look at how the weight of a fish and its fertility depends on its age. Similarly you can make a plot or a *bar graph* of the current numbers of fish versus age. (Type `help bar`.)

A first attempt at a Leslie matrix (cf. Chapter 2) has been provided. Type `help lmfish` to see what it is supposed to do. To see the whole M-file in more detail use Notepad. You will find that several additions and corrections are required, for example to take account of the fact that the egg-production (fertility) rate refers only to *female* fish and to model more realistically the removal rate due to fishing. The latter clearly depends on the mesh size of the fishing nets, the size distribution of the particular fish species and the frequency of harvesting. Part of this procedure will require you to establish a mathematical model which fits the weight *vs* age data and also predicts how the cross-section of a fish depends on its age.

We will return to each of these tasks in due course. In the meantime, it is worth trying the first two exercises at the end of the chapter.

20.2 Models of fish growth

There are two simple models which you are invited to consider. Further details of these can be found in [3]. The predictions of each should be compared with the given data on fish weight as a function of age. First, we define a few variables and parameters which will come in handy:

$m(t)$	–	mass of a fish at time t
$V(t)$	–	volume of a fish at time t
t	–	age of the fish in years
ρ	–	density of a fish in g/cm^3 (pure water is 1 g/cm^3)
h, w, l	–	height, width and length of a fish (cm)

(*Note*: in colloquial usage, the terms *mass* and *weight* are often used interchangeably. Actually the weight and mass are numerically the same only if the former is expressed in units of g.)

You should be able to estimate a value for ρ simply from the fact that the fish swim in water with no apparent difficulty. Specifying the dimensions h, w and l pre-supposes a very simple rectangular box model of a fish. You may be able to do better. The box model implies that the volume $V = hlw$. In any case, $m = \rho V$. We will return to these points.

20.2.1 Gompertz growth model

In this model, the rate of growth (increase in mass) is proportional to the amount of tissue doing the growing but with the factor of proportionality decreasing exponentially with time. This models an aging process and gives

$$\frac{dm}{dt} = re^{-\lambda t} m \,, \tag{20.1}$$

where the parameters r and λ (both > 0) have to be determined to fit the circumstances. Clearly r sets the overall rate of growth and λ determines the time-scale over which the period of rapid growth takes place. You can estimate $1/\lambda$ for our fish very roughly by looking at your plot of weight gain to see at what age the maximum rate of growth occurs. Likewise you can estimate r very roughly by applying the formula (20.1) at a couple of t-values, where m and its slope can be estimated. Make a rough guess of each in this way and use fodesol to see how the model behaves. Note that you have to supply an initial condition. Choose a t-value for which you know the corresponding m-value ($t = 1$?). You can adjust the parameters in the light of what you see.

However, you might well find you can solve the ODE (20.1) by integration to obtain an analytic solution for $m(t)$. A good plan would be to do this and then use fodesol with suitably modified fnxt to obtain a plot and numerical solution as a cross check. In this way you can be sure of correct answers. If unsure of fodesol, you can have another look at Chapter 7.

20.2.2 Von Bertalanffy growth model

This more sophisticated model, here written for the increase in volume rather than mass, gives

$$\frac{dV}{dt} = \alpha V^{\frac{2}{3}} - \beta V \,, \tag{20.2}$$

where the parameters α and β are positive. The idea behind it is one of energetic balance. The first positive term represents energetic (nutrient) input and is proportional to surface area (hence $V^{\frac{2}{3}}$), while the negative second term represents the drain on resources, that is, food for body tissue, and is proportional to the volume itself. Again, you should try to solve this both graphically/numerically with `fodesol` and analytically where you may find the change of variable $L = V^{\frac{1}{3}}$ helpful. Here, L obviously has the dimensions of length. To estimate the parameters in this case think about the maximum size achieved when $dV/dt = 0$. You can also look at spot values of the slope of the data if this helps. (Use two neighbouring points.)

20.2.3 Choosing the best model and parameters

Using the above methods you should be able to arrive at a reasonable set of parameters to describe the fish growth data [`fish(:,2)` vs `fish(:,1)`]. If you actually manage to obtain correct analytic solutions, you can go much further and try a 'least squares' best fit of the data. This will show you which growth model is better and also determine the best fit parameters for it.

The idea of least squares was discussed in Chapter 5 within the context of linear least squares. The same idea of minimising the residual function R^2 with respect to the choice of parameters can be applied to any parametrisation provided you have a means of minimising. MATLAB is able to do this (of course) and appropriate M-files to perform a multi-parameter fit `mparst4.m` and `mparft.m` have been set up to do this for you. Type `mparst4` to set up the example needed here, followed by `mparft`. This particular M-file sets up a fit of the mathematical model

$$m = ae^{b\sqrt{t}} + c \qquad\qquad (20.3)$$

to the growth data. When prompted, try giving [1 1 1] as suggested starting values for the parameters [$a\,b\,c$]. As you will see from the final plot and quoted residual R^2, the fit is rather poor! You could now modify the fitted function in `resid4.m` to reproduce, in turn, your two model parametrisations (solutions to equations (20.1) and (20.2)).

The best parametrisation can be identified either by eye, if the difference is marked, or by looking for that with the smallest residual R^2 which measures the sum of the squares of the differences between the model and the data.

Now try the exercises concerning the growth models and their parameters.

20.3 Designing the Leslie matrix

You may need to review the structure and use of Leslie matrices as illustrated in Chapter 2. The M-file `leslie.m` is a simple example. The file `lmfish.m` is provided as a starting point for further development according to your needs. It already incorporates the following features:

(i) It uses the egg-production data in `fishy.m` and the additional information that only 1 in 5000 eggs survive to give a one year old fish. However it needs altering to take account of the fact that the egg-production rate supplied by `fishy.m` is per *female* fish. You should correct that point now.

(ii) It also incorporates a non-survival probability, different for each age group. The probabilities in here are just more or less arbitrary numbers (less than one of course) and increasing since we expect older/larger fish to be more likely to be caught by a fishing net.

(iii) It has the key feature of a Leslie matrix that the survivors from one age band grow older and so populate the next age band.

(iv) No fish survives beyond 10 years of age!

These features are annotated in the M-file. Your next task should be to improve the survival probability model to incorporate your knowledge of the size of a fish of a given age and to take account of different mesh sizes.

20.3.1 Cross-section of a fish

At this point you may have a good model which tells you the mass and hence volume V at any arbitrary time t. If you didn't succeed in getting a very reliable model for that, you can still proceed by using the data points `fish(:,2)` themselves (modified by the density if necessary) since you will normally only need $V(t)$ at integer values of t anyway.

To estimate how easily the fish can go through a net of mesh size d cm, say, you need to relate V to the transverse dimensions of the fish. The simplest possible model is to approximate the fish by a rectangular cuboid. You then need to make some assumption about the shape, that is, ratio of length to height (l/h) and width to height (w/h). From this you can deduce h, say, from V. Clearly, if $h > d$ the fish is caught. You may have a better model for the fish shape. If so, try to establish a formula for its maximum 'height' in terms of its volume as shown above.

20.3.2 *Probability of getting caught*

Even if $h < d$, the fish can become snarled up in a moving net. The best idea is to dream up a formula for the probability that a fish of height h will get caught in a net of mesh size d. For example, the probabilty function

$$P_C = \min\{1, (h/d)^2\} \tag{20.4}$$

means certain capture for fish bigger than the mesh size and a decreasing likelihood for smaller fish. One might justify using $(h/d)^2$ rather than some other function by claiming that the cross-sectional *area* rather than a linear dimension is significant here. There is a further modelling feature which you should include and that is the efficiency of trawling (harvesting) in any one year. Even if a very fine mesh is used, the total catch depends on how often and how completely this is done. You could combine these factors into one 'efficiency' factor f_{eff} with which to multiply P_C before going on to calculating the probability $1 - P$ of *not* getting caught in a given year.

You should adapt `lmfish.m` to incorporate as many of the above features that you can. The appropriate places to insert these additions are annotated in the M-file. Add your own comments (lines beginning with %) as you add more features. This helps to remind you what you have done. Since you will want to use this code repeatedly to obtain and study the Leslie matrix H, it might be as well to get the M-file to *prompt* you for the necessary values of d and f_{eff}. You can do this with a line such as

```
d=input(' Give me the mesh size d ');
```

which will do just that and make the input value of d (10 say) available to subsequent lines in `lmfish.m`. In this way you can keep reusing the M-file without having to edit it every time you need a new value!

20.4 Fishing strategy

If you understood the Chapter 2 material on Leslie matrices very well, you will recall that the eigenvalues of the matrix determine the long term fate of a population. If the largest is greater than 1, it will prosper. If it is less than 1, the population will eventually die out. This gives you a quick way of determining whether harvesting each year (multiplying a population vector by H) will eventually wipe out the stocks of fish. You can check explicitly what happens in the first year by taking the current population, for example

```
>> p0=fish(:,3)
```

and performing one year's fishing by forming

```
>> p1=H*p0
```

and so on forming [p0 p1 p2 p3...]. You might like to display these on a single plot to get an idea of what is really happening. You will find that the fate of the fish stocks depends crucially on the net size d and the frequency of fishing in the year f_{eff}. There are several possibilities for establishing a sensible fishing policy. You can keep fishing at a constant low level with a sensibly chosen net size so that the younger fish have a chance to grow. Alternatively you could conduct heavy fishing for a year or so then give the stocks a rest. You could model this by having two different Leslie matrices H_1 and and H_0 which are multiplied in different sequences according to your strategy. Again, you can check the overall fate of the population by looking at the eigenvalues of the products.

There are many things to experiment with. The fishing industry is not directly interested in the numbers of fish swimming about, but in the maximum biomass (total mass of fish) which can be removed over an extended period. Ideally, your strategy should be to maximise this yield. You should work out how to extract the numbers caught each year in each age group. Perhaps the simplest way is to adapt a version of lmfish.m to produce a diagonal matrix which, when applied to the population of a given year, gives a vector of caught fish. Multiplying (.*) by the vector of fish weights gives the vector of biomass caught.

There are also many imperfections in such a model which you should think about. For example, if you set $d = 100$ and $f_{eff} = 0$ the model suggests an exponentially growing population of fish. Of course, this does not actually happen since the food supply is finite and disease can set in. If the population is kept under control, however, one may be able to ignore these limitations.

Exercises

20.1   Use the data loaded in with fishy to construct a bar graph of the total mass of fish (the biomass) as a function of age.

20.2   Find the total biomass (in some sensible units) of fish of all ages currently swimming about in one square kilometre of sea.

20.3 Estimate the fish density ρ (in g/cm^3) and hence obtain a plot of the height, width and length of the fish as a function of age according to the rectangular 'box' model of a fish.

20.4 Solve equations (20.1) and (20.2) by the method of separation of variables.

20.5 Use `fodesol` to obtain numerical solutions of equations (20.1) and (20.2) with guessed values of the parameters. By comparing with the data for fish mass (or volume), try to find better values of the parameters.

20.6 If you managed to obtain (by integration) closed expressions for $V(t)$ for either of the models, use `mparft` to obtain 'best fit' parameters and compare these with those which you estimated with the help of `fodesol`.

20.7 Give your estimate of the parameters $[m_0, r, \lambda]$ or $[m_0, \alpha, \beta]$ of the model (20.1) or (20.2) which, in your opinion, best describes the growth data for fish. Also give an estimate of the maximum weight and length which one of your model fish could achieve.

20.8 Estimate how the cross-section of a fish varies with age.

20.9 Adapt the Leslie matrix `lmfish.m` to incorporate the effects of net size, fish size and fishing efficiency as discussed above.

20.10 Give an example of a viable fishing strategy (net size and efficiency factor which will yield a good harvest but conserve stocks after five consecutive years fishing. What annual biomass does your harvest yield from each square kilometre?

20.11 Comment on alternative strategies in which years of heavy fishing are interspersed with recovery years. Obtain some numbers or plots to support these comments.

21

Epidemics

Aims of the project

You are given some data describing the development of epidemics which occurred in various communities and are provided with a basic model describing the dynamics of such a system. You are then invited to analyse the data as best you can to discover the underlying behaviour of the disease and the response of the community to it. The model is based on a set of coupled first order differential equations. You must obtain approximate analytic solutions and full numerical solutions using the routines provided.

Mathematical ideas used

You will need to: work with coupled first order differential equations (Chapter 7). make linear approximations; know how to integrate simple linear first order equations; understand the least squares fit idea (Chapter 5);

MATLAB techniques used

The numerical techniques for differential equations are those first introduced in Chapter 7. The multi-parameter least squares fit package is that first described in Chapter 20.

For convenience, here is a list of relevant M-files, both standard ones and special ones provided for this particular project. You may have to modify some of these in the course of your study. In each case typing `help` will give information on the purpose and usage.

fludat.m	–	data for school flu epidemic
plagdat.m	–	data for Bombay plague
colddat.m	–	data for island cold epidemic
sirepi.m	–	SIR epidemic model integrator
sirfn.m	–	derivative function for the above
mparft.m	–	multi-parameter least squares fit
mparst3.m	–	example set up for the above
resid3.m	–	residuals function for the above
lagsum.m	–	cumulative sum

21.1 Preliminary look at some data

Use the command `fludat` to load some sample data, in this case the number of boys infected with flu as a function of time in days during an outbreak of the illness in a public school. You can regard the school as a closed community of N (= 763) individuals with initially perhaps just one infected pupil ($I = 1$) and $S = N - 1 = 762$ potentially 'susceptible' individuals. As time progresses, the number I of infected pupils rises but not indefinitely because there are fewer left still susceptible to infection. The ones who have successfully recovered have now got some, at least temporary, immunity and so can be thought of as 'removed' from the system (R). This classification of the N pupils into these three classes of individuals

$$N = S(t) + I(t) + R(t) \tag{21.1}$$

is used in the so-called SIR model described in the next section.

The M-file `plagdat.m` also loads some data, this time for the number of deaths per week report during a plague epidemic in Bombay in 1905 and 1906. Unlike the above case, the class of 'removed' individuals $R(t)$ is unfortunately due to death rather than acquired immunity. The net result from the point of view of model building is, however, the same: they take no further part in receiving or promoting infection. The datum loaded is, roughly speaking, the *rate* of removal (death), that is dR/dt.

The third data sample `colddat` is similar to the first two data samples and concerns the spread of a common cold outbreak within the population of a remote Atlantic island following one of the infrequent visits of a supply ship. This time the numbers of new cold cases each day were recorded. The total island population was 280.

21.2 The SIR model for the dynamics of an epidemic

This model, due originally to Kermack & McKendrick in the 1930s, uses first order coupled differential equations to describe the evolution of the populations S, I and R identified as above, hence the model's name. For an introduction to such models of epidemics see, for example, [10] or [3]. Since the three variables sum to a constant (see equation (21.1)), only two of the following equations are actually independent:

$$\frac{dS}{dt} = -rSI, \tag{21.2}$$

$$\frac{dI}{dt} = rSI - aI, \tag{21.3}$$

$$\frac{dR}{dt} = aI. \tag{21.4}$$

To recap:

- $S(t)$ is the number of 'susceptibles', i.e. the number in the community of N ready and waiting to be infected with the disease at time t.
- $I(t)$ is the number who actually have the disease and are capable of infecting others (the susceptibles) and are no longer susceptible themselves.
- $R(t)$ is the number 'removed' from the system because they are neither infected nor susceptible since they have either recovered, and so acquired immunity, or, in severe cases, are dead.

The origin of the simple model represented by the right-hand side of these equations is very simple to understand in the light of the above examples. Look at equation (21.4). This says that the rate at which people are removed (die or recover) is proportional to the number of cases I. Here

$$a = removal\ rate\ of\ infectives \quad (a > 0). \tag{21.5}$$

The second term in equation (21.3) reflects the corollary to this statement since these same individuals are being *lost* from class I. The first term in this equation is the statement that the number of new cases per unit time is proportional both to the number of potential victims (susceptibles) S and to the number already having the disease I and so capable of infecting others. The constant of proportionality is

$$r = infection\ rate \quad (r > 0). \tag{21.6}$$

Finally, equation (21.2) represents the corollary to this statement since these same individuals are being lost to class S.

Any particular model is now fully specified by the infection and removal rates a and r and by a set of intial conditions. Usually, one takes $R(0) = 0$ since no one has yet had the chance to recover or die. Thus, a choice of $I(0) = I_0$ is enough to define the system at $t = 0$ since then $S_0 = N - I_0$. Typically, one takes $t = 0$ to be the moment at which $I = I_0 = 1$ if the epidemic is triggered by a single infected individual.

21.3 Studying the behaviour analytically

The first thing to notice is that $dS/dt < 0$ at all times so that $S(t)$ is monotonically decreasing whereas

$$dI/dt = 0 \quad \text{when} \quad S = \rho \tag{21.7}$$

and so $I(t)$ can have a stationary point at some time $t_{\max}$. This is easily seen to be a maximum since I(t) is usually rather small initially. Here

$$\rho = a/r \tag{21.8}$$

is called the *relative removal rate*. Thus, if $S_0 < \rho$, condition (21.7) can never be satisfied and the infection dies out. There is no epidemic. The size of ρ is therefore a key parameter in the model. The definition of an epidemic is just that $I(t) > I_0$ at some later time t.

One can partially solve the system by eliminating t as an explicit variable and solving the resulting differential equation. For example, from equations (21.2) and (21.3)

$$\frac{dI}{dS} = -1 + \frac{\rho}{S}, \tag{21.9}$$

which you should be able to show gives the solution

$$\frac{R}{\rho} = \ln\left(\frac{S_0}{S}\right) \tag{21.10}$$

In general, you still need to solve the system numerically to get the t dependence. However, in certain circumstances, you can make approximations to progress further analytically. For example, if you put the exact result (21.10) back into the differential equation for $R(t)$ (21.4), you get

$$\frac{dR}{dt} = a(N - R - S_0 e^{-\frac{R}{\rho}}). \tag{21.11}$$

As it stands, this requires numerical solution. For example, we could use `fodesol.m` and the techniques of Chapter 7. However, when $R/\rho \ll 1$ the exponential can be expanded to, say, second order in R giving a separable variable equation which is tractable. To do the integration, one can use 'completing the square' to get something of the form

$$\frac{dR}{A^2 - (R - \bar{R})^2} = \sigma dt. \tag{21.12}$$

You should try this for yourself (see the exercises). The result is an explicit function $\tilde{R}(t)$ which is then an approximate solution to equation (21.11).

If you then calculate $d\tilde{R}/dt$ you get an expression which could, for example, be used to compare directly with the plague data loaded with `plagdat.m`. The result is of the form

$$\frac{d\tilde{R}}{dt} = A\alpha \operatorname{sech}^2(\alpha t + \beta). \tag{21.13}$$

Equation (21.10) can also be used to study the final outcome of the epidemic, when I becomes effectively zero as $t \to \infty$. In that case, equation (21.10) can be rewritten

$$S(\infty)/S_0 = \exp\left(-\frac{(N - S(\infty))}{\rho}\right) \tag{21.14}$$

which could be solved numerically, for example using the MATLAB function `fzero`. If we happen to know the final number of susceptibles and the total community involved, we can use this to estimate ρ. Equation (21.10) can also be combined with equation (21.7) to get further relations between the parameters and the height of the maximum epidemic $I_{\max}$. There are lots of different ways of using such relations depending on the type of data available.

21.4 Analysing the data

In the case of the plague data, you don't know much about the model parameters – even the size of the relevant Bombay community. However, assuming that R/ρ is not too big, you can test the model via its prediction (21.13) by trying to find parameters A, α and β which fit the data in the sense of a least squares fit. The least squares fit described in Chapter 5 involved a straight line fit. Here, and in Chapter 20, the parametrisation is more complicated but the same principle applies: the best fit is one which *minimises* the sum of the squared 'residuals' between theory and data. The M-file `mparft.m` does this. You should type `mparst3` to set up the example provided which attempts to fit a parametrisation

$$\frac{dR}{dt} = Ke^{-\frac{(t-t_0)^2}{2b^2}} \tag{21.15}$$

to the plague data. You then type `mparft` to start the least squares fit and follow the prompts. The 'residuals' are calculated in an M-file called `resid3.m`. When you have tried this successfully, you should go on to modify a copy of `resid3.m` to implement the model expression (21.13).

A full numerical solution of the SIR model is accomplished using the M-file `sirepi.m`. You should try to find parameters a and r which give a reasonable description of the school flu data. In order to get rough estimates of the sort of values to try, you can study the small t and large t data separately. At small t, before R becomes large you can approximate S by $N - I$ and equation (21.3) by

$$\frac{dI}{dt} = rNI. \tag{21.16}$$

By solving this and comparing the result with the small t data, you can get an estimate of r. Similarly, at large t when the epidemic has run its course, S and I are both small so that dI/dt is dominated by the last term. Again, you can integrate this equation and get an estimate of a this time. You can use these rough values to get started with `sirepi`.

In the case of the Atlantic island data, `colddat.m` provides you with the number of *new* cases each day which is clearly something to do with the infection rate. To get the actual number of infectives $I(t)$, you need to know something about how long the infection lasts. This will vary but a reasonable model might be to take it to be, say, 7 days. From this you can deduce, not only $I(t)$, but $R(t)$ and hence $S(t)$ as well. You can manipulate the data by hand or you can use MATLAB's facilities to help you. An M-file `lagsum.m` is provided which may be of some use. Given a complete breakdown of the population into the categories S, I and R, you can test the SIR model in some detail. By comparing with the SIR model predictions you can decide whether your cold infection period of 7 days was too long or too short and so on.

Exercises

21.1   Check the nature of any turning value given by equation (21.7) using $\ddot{I}(t_{max})$.

21.2   Derive equation (21.10) from equation (21.9) using the initial conditions discussed at the end of §21.2.

21.3   Integrate equation (21.12) using the standard integral

$$\int \frac{1}{a^2 - x^2} = \frac{1}{a} \tanh^{-1}\left(\frac{x}{a}\right) \qquad (21.17)$$

to obtain equation (21.13) and try to find expressions for A, α and β in terms of the model parameters a, r, N and S_0.

21.4   Find parameters which give a reasonable description of the school flu epidemic data assuming that $I = 1$ at $t = 0$ and the total number of boys in the school is $N = 763$.

21.5   Estimate R/ρ at the height of the flu epidemic?

21.6   How many boys have escaped the flu by the fifteenth day?

21.7   Find the parameters A, α and β which best describe the plague data using the approximate expression (21.13).

21.8   Adapt a copy of `sirepi.m` to include a plot dR/dt and so try to find a complete model (N, r and a) for the plague data. You will have to make some choice of the total community involved in the Bombay plague.

21.9   Compare the parameters found in this model with those from the best fit expression found previously.

21.10   Convert `colddat.m` to supply a table and plot of estimates of $S(t)$, $I(t)$ and $R(t)$ for the Atlantic island cold data. Estimate $R(\infty)$.

21.11   Find reasonable SIR parameters to describe the Atlantic island cold data. Comment on the success or otherwise of your modelling attempt.

22

Dynamics of Snowboating

Aims of the project

You are given a mathematical description of a winter sports area and invited to help design a run for the new winter sport of 'snowboating' – sliding down mountainsides in a rubber boat, more or less out of control. You will have to obtain reasonable model parameters to define safe operating conditions.

Mathematical ideas used

You are given a fairly realistic dynamical model based on Newton's laws. The equations of motion are coupled differential equations.

MATLAB techniques used

You can solve the differential equations using the numerical methods and MATLAB programmes of (Chapter 7). For convenience, there follows a list of relevant M-files, both standard ones and special ones provided for this particular project. You will have to modify some of these in the course of your study. In each case typing `help` gives information on the purpose and usage.

`topog`	–	3D and contour plots of slopes
`snowsl`	–	definition of the topographical surface
`snowboat`	–	solves the equations of motion for the snowboat
`snbtfn`	–	derivative function for the above
`fsnow0`	–	surface function f (scalar args.)
`fsnow1`	–	f_x and f_y
`fsnow2`	–	f_{xx}, f_{xy} and f_{yy}
`snowmn`	–	minimisation function for the surface
`snowmx`	–	maximisation function for the surface

22.1 Preliminary look at the problem

Use the command `topog` to show a mesh plot followed by a contour plot
of the snow hills where the action will take place. It is defined by

$$z = f(x,y), \tag{22.1}$$

where f is some smooth function which can be examined in either
`snowsl.m` or `fsnow0.m`. The M-file `topog.m` also allows you to find
out where the hill tops (maxima) and bottoms of the valleys (minima)
are. It uses the standard M-file `fmins` to minimise $-f$ and f to find
local turning points. Actually, you are provided with the first and second order partial derivatives (see `fsnow1.m` and `fsnow2.m`) so you could
use these to find the turning points analytically. However, if you look
inside them, the derivatives are a bit long and messy and so the general
numerical minimisation routine `fmins` is more convenient.

Your first task should be to find and note down some of the geo-
graphic features. In the absence of any frictional forces, a snowboat (an
unsteered toboggan) set going from the highest point will head down
the path of steepest descent gathering speed continually. Since in that
case the total energy is conserved, the speed v at any moment can be
calculated from the loss in potential energy:

$$E = E_P + E_K = \text{constant}, \tag{22.2}$$

where

$$E_P = mgz, \quad E_K = \frac{1}{2}mv^2. \tag{22.3}$$

You can work out what speed the snowboat will be doing when it
drops to its lowest point – a totally unrealistic and dangerous speed. Of
course, in real life, there are resistive forces:

- friction between boat and snow, which will be more or less constant;
- air resistance which rises with speed.

Part of your task will be to obtain reasonable values for the parameters
describing these forces. If necessary, you can increase the friction by
roughening the underside of the boat and so obtain a slower ride. The
other major task will be to find a safe starting point and initial velocity
to give an exhilarating, but not life-threatening, ride down the mountain.

Before continuing, try the first three exercises at the end of the chapter. These will help you visualise the problem.

22.2 The equations of motion

First, we define some key constants and variables.

m –   mass of the snowboat: take as 150 kg including one passenger;

A –   cross-sectional area: say 1.5×0.5 m^2;

g –   acceleration due to gravity: 9.81 m/s;

μ –   coefficient of sliding friction between boat and snow;

k –   coefficient of air resistance ($F_k = kv^2$);

$\mathbf{r}$ –   position vector of snowboat;

$\mathbf{F}$ –   total external force on the snowboat;

$\mathbf{F}_g$ –   gravitational force;

$\mathbf{F}_\mu$ –   resistive force due to sliding friction;

$\mathbf{F}_k$ –   force due to air resistance;

$\mathbf{N}$ –   normal reaction from the snow surface.

According to Newton's second law,

$$m\ddot{\mathbf{r}} = \mathbf{F} = \mathbf{F}_g + \mathbf{F}_\mu + \mathbf{F}_k + \mathbf{N}, \qquad (22.4)$$

where the forces acting are shown schematically in Figure 22.1.

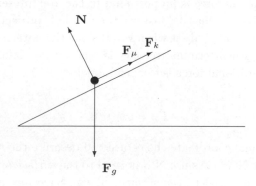

Fig. 22.1. Forces acting on the snowboat (not to scale).

While the snowboat remains on the slope, the motion is described by equation (22.4) together with equation (22.1). If it goes over a bump or cliff and leaves the ground, it will behave as a projectile in a resisting medium subject only to equation (22.4) with $\mathbf{F}_\mu = \mathbf{N} = 0$. For sliding, the following is a reasonable model of friction

$$\mathbf{F}_\mu = -\frac{\mu \dot{\mathbf{r}}}{|\dot{\mathbf{r}}|}|\mathbf{N}|. \tag{22.5}$$

Notice that friction always acts in the opposite direction to the sliding. This simple model therefore breaks down when $\dot{\mathbf{r}} = \mathbf{0}$ where the force changes *discontinuously*! The numerical solution method described in Chapter 7 could have problems with this. Fortunately, we are not interested in following the snowboat *after* it has come to rest ($\dot{\mathbf{r}} = \mathbf{0}$)! It will be left to you to determine a realistic value for μ. (See the exercises.) For the air resistance, it is found that cars, people on bicycles, free-fall parachutists etc. experience a force like

$$\mathbf{F}_k = -k\dot{\mathbf{r}}|\dot{\mathbf{r}}|. \tag{22.6}$$

Again, this always opposes motion but has reasonable behaviour as the velocity goes to zero. You should be able to figure out a sensible value for k using the experimental information that a free-fall parachutist weighing the equivalent of 70 kg, in a spreadeagled pose, reaches a terminal velocity of around 54 m/s. To convert his/her k-value to that for a snowboat, you will need to estimate the frontal area of each and use the fact that the force is proportional to the area presented. To get the terminal velocity equation, just use the free-fall (projectile) equation described above and study it when $\ddot{\mathbf{r}} = \mathbf{0}$ since the parachutist is no longer accelerating when terminal velocity is reached (by definition!)

The gravitational force is easy to model:

$$\mathbf{F}_g = mg(0,0,-1) \tag{22.7}$$

using the same coordinate choice used to describe the snowfield.

Note that $\mathbf{N} \cdot \dot{\mathbf{r}} = 0$ since $\mathbf{N}$ is normal to the surface and $\dot{\mathbf{r}}$, the velocity vector, is tangential to the surface. If you have met some elementary vector calculus, you may know that the normal to the surface (22.1) is parallel to the **grad** of the function $\phi = z - f(x,y)$. This gives an easy way of finding it numerically.

Newton's equations (22.4) therefore represent a set of three coupled second order differential equations where only *two* coordinates (x and y say) are independent because of (22.1). If we eliminate z, $\dot{z}$ and $\ddot{z}$ using this, we get two second order coupled equations in x and y. Using the trick learned in Chapter 7, these can be reexpressed in terms of *four* first order equations in x, y, u and v, say, where

$$u = \dot{x}, \quad v = \dot{y}.$$

The first order equations are

$$\begin{pmatrix} \dot{x} \\ \dot{y} \end{pmatrix} = \begin{pmatrix} u \\ v \end{pmatrix}, \quad \begin{pmatrix} \dot{u} \\ \dot{v} \end{pmatrix} = A^{-1} \begin{pmatrix} -\frac{k}{m}|\dot{\mathbf{r}}|u + q_x\tilde{g} \\ -\frac{k}{m}|\dot{\mathbf{r}}|v + q_y\tilde{g} \end{pmatrix}, \quad (22.8)$$

where

$$\mathbf{A} = \begin{pmatrix} 1 - q_x f_x & -q_x f_y \\ -q_y f_x & 1 - q_y f_y \end{pmatrix}.$$

Here,

$$\tilde{g} = g + f_{xx}u^2 + 2f_{xy}uv + f_{yy}v^2 + \frac{k}{m}|\dot{\mathbf{r}}|(f_x u + f_y v),$$

$$q_x \equiv \frac{p_x}{p_z}, \quad p_x = n_x - \frac{\mu\dot{x}}{|\dot{\mathbf{r}}|}$$

and (n_x, n_y, n_z) is parallel to **grad** ϕ, the surface normal.

You may wish to verify these equations but be ready to do quite a bit of calculus and algebra.

The M-file snowboat.m solves these equations numerically in the same manner as the examples studied in Chapter 7. The function snbtfn.m simply contains the derivatives as expressed in equation (22.8).

22.3 Exploring the operating parameters

Having found out where the hill tops are, made a reasonable estimate of the air resistance parameter k and guessed a trial value for μ (the coefficient of sliding friction.), you can use snowboat to check out a sensible run. Notice that if you start on a relatively flat area you will need quite a large downhill push (large $\sqrt{u_0^2 + v_0^2}$) to get going. Remember that the friction model is not very reliable at low speeds on gently sloping ground where the snowboat is liable to grind to a halt.

A useful way to proceed is to choose the surface function f in equation (22.1) to be that for a simple sloping plane. In this case, you can solve the equations analytically and check out the performance of the numerical procedure. To do this, make *new copies* of snowsl.m, fsnows0.m, fsnows1.m, and fsnows2.m corresponding to the choice of plane, say

$$z = f(x, y) = y/2. \quad (22.9)$$

The partial derivatives are not hard to work out! The problem becomes one dimensional since everything can be written in terms of, say, y. To solve the single second order ODE, the best method is to use v and y rather than y and t as the dependent and independent variables respectively, using

$$\ddot{y} = v\frac{dv}{dy} \, . \tag{22.10}$$

In this way, you can check the validity of much of the formalism and see how well the numerical procedure is performing.

The M-file `snowboat.m` prompts you for the parameters μ and k, the starting velocity (horizontal components) u_0 and v_0 and a time interval δt over which to integrate numerically. Note that you control the starting direction with u_0 and v_0 making them negative if necessary. The subsequent direction of motion is determined by the topographical features of the snow slopes via the equations of motion (22.8). Ideally, you want the run to start near the top of a hill and grind to a halt near the lowest point within the area. Excursions into neighbouring valleys are expensive.

Exercises

22.1   Print out, or make rough sketch copies of the 3D-view plot and of the contour plot provided by `topog`.

22.2   Give the locations of all hill tops and bottoms of valleys within the area which is $2\,\text{km} \times 2\,\text{km}$. Give the altitudes as well.

22.3   Sketch on the plots a couple of paths suitable for snowboating.

22.4   Quote your estimate of a reasonable value for k based on the free-fall parachutist data and your estimate of the relative frontal areas involved.

22.5   Modify the surface function routines to describe the sloping plane (22.9) as discussed above. Use `snowboat` to describe motion down the plane from the point $(0, 2000)$ with initial velocity $(0, -2)$ m/s and compare your results with the exact answers got by integrating the equation analytically.

22.6   If, instead, the sloping plane had the equation $z = y/10$ what value of μ, the coefficient of friction between snowboat and snow, would lead to zero acceleration down the hill?

22.7   Find a reasonable coefficient of friction μ, and initial velocity which allows the snowboat to reach the bottom of the valley reasonably smoothly starting from the very highest point in the snow hills (use the original versions of `snows1.m`, etc.). Give the initial parameters, the approximate time taken and the approximate top speed reached during the descent.

22.8   Try to find other similar runs from the remaining hill tops. Comment on how the initial velocity might be achieved in each case.

22.9   Modify `snowboat.m` to display the change in potential energy and the change in kinetic energy from the initial and final positions and give the results for the trip down from the highest hill.

22.10   Give the energy differences for the same run but when the friction and air resistance are switched off. Comment on the two sets of results.

23

Tides

Aims of the project

You are given data consisting of a series of tidal measurements taken over some period of time and are required to analyse these, draw some conclusions and make some predictions. Specifically, the data are the measured heights (in metres) of the sea level taken at the given equally spaced times (in hours) at a fixed place. You are asked to find such things as the mean sea level, mean tidal range, mean period between high tides, estimated sea level at particular time on a particular day etc. The times given, are with respect to $0:00$ hours on 1 January 1992.

Mathematical ideas used

Since tidal effects are clearly periodic in nature, you are given some assistance in learning about Fourier series and their application.

MATLAB techniques used

There are a number of built-in MATLAB routines which can be used to help with Fourier analysis and manipulating vectors of data:

fourier	–	demo of Fourier series for square wave;
fftdemo	–	demo of FFT for signal analysis;
fft	–	Fourier transform;
ifft	–	inverse Fourier transform;
spline	–	interpolation of data points;
mean	–	mean of data;
max	–	maximum of data.

In addition to these standard procedures, the following M-files are provided for this project. You may have to modify some of these in the course of your study. In each case typing `help` gives some information on the purpose and usage.

tides.m	–	load tidal data tm(1:nt), hm(1:nt);
fourdat.m	–	load sample signal data;
fouran.m	–	Fourier analysis of data;
foursig.m	–	for checking purposes only;
locmax.m	–	local maxima of data;
times.mat	–	vector of tide measurement times;
heights.mat	–	vector of tide heights;
noisyt.mat	–	vector of sample signal times;
noisys.mat	–	vector of sample signal data.

23.1 Preliminary look at the tidal data

Use the command `tides` to load the tidal information from the MAT-LAB data files `times.mat` and `heights.mat`. Now have a look at the data. The times (`tm`) are in hours and the heights (`hm`) are in metres. The first few exercises at the end of the chapter should help you to get a general picture of what is going on.

Having explored these and other features, you will see that the tidal heights very roughly satisfy something like

$$h(t) = h_0 + A\sin(\omega t + \phi). \tag{23.1}$$

Estimate h_0, A, ω and ϕ as best you can and comment on the reliability of each of these estimates. To do this, start by estimating A by noting the 'amplitude' of the main oscillation which is half the difference between maximum and minimum. Then establish ϕ by noting where each the oscillation period starts. You can estimate h_0 from the water level at mid-tide.

23.2 Fourier series and methods

If the tides exhibited an exact sinusoidal behaviour as in equation (23.1), there would be no need for tide tables. In fact, the amplitude A and the frequency ω defined in this way vary with time. Another, more precise, way to describe this is to recognise that more than one frequency and corresponding amplitude are present. In general, any periodic disturbance such as a sound will contain many modes of oscillation. For an orchestra this is obvious. However, even the purest note of a top class singer contains many frequencies besides the 'fundamental'. You may well have seen demonstrations of this using an oscilloscope to analyse and display the frequency content of a human voice or other sound.

Mathematically, the total sound or other signal $f(t)$ is decomposed as a Fourier sum (discrete time sample version) or Fourier integral (continuous time version) over its constituent modes. The Fourier sum looks like

$$f(t) = \sum_{n=-\infty}^{\infty} A_n e^{\frac{2\pi int}{T}}, \qquad (23.2)$$

where T is the total time interval over which measurements are made and n runs over all integers. You sometimes see this written in terms of $\sin$ and $\cos$ since these are related to the complex exponential:

$$e^{\pm i \frac{2\pi nt}{T}} = \cos\left(\frac{2\pi nt}{T}\right) \pm i \sin\left(\frac{2\pi nt}{T}\right). \qquad (23.3)$$

Since $\sin(2\pi nt/T)$ and $\cos(2\pi nt/T)$ have period T/n as a function of t, each mode labelled by n corresponds to frequency n/T. Conversely you can write the frequency coefficient A_n as an integral over t

$$A_n = \frac{1}{T} \int_0^T f(t) e^{-i \frac{2\pi nt}{T}} dt. \qquad (23.4)$$

Where the signal $f(t)$ is only known at discrete (but regularly spaced) times, this *integral* can be replaced by a *sum*. This is essentially what the MATLAB routines `fft` (fast fourier transform) and its inverse `ifft` do. There are two useful MATLAB demos to help you understand how Fourier sums and transforms work.

- **fourier**: this shows how a periodic 'square wave' can be reproduced by a (infinite) sum of sinusoidal frequencies. Because the shape of the periodic signal is far from sinusoidal here, you need many sine waves to reproduce the original signal well. Actually, when discontinuities are present, the Fourier sum inevitably gives the wrong limit at the point of discontinuity – the so-called 'Gibbs phenomenon'.

- **fftdemo**: this demo starts by making up a signal which is basically the sum of two sinusoidal ones (that is, two frequencies) but adds some random noise to make it look like the sort of thing you might encounter playing music over a noisy telephone line. The data look very random when plotted against time (much more so than the tide data!). However, when you look at the frequency coefficients (Fourier transform of the data) you clearly see the original sinusoidal components. Note that the frequency unit 1 Hertz is just 1 s^{-1}.

It is clear that a similar Fourier analysis will help one understand what influences the tidal data. Which periods or frequencies dominate?

23.3 Analysis of an electrical signal

As a prelude to analysing the tidal information, some data similar to that in the fftdemo have been provided (fourdat) together with a basic analysis M-file (fouran). As usual, typing help fname will give you some help in getting started. You should identify which frequencies are present in the original signal data and estimate the relative importance of each frequency component. To get this, look at the square of the coefficient of the mode as given by the Fourier transform. You may wish to play around with a copy of fouran.m to get it to display or plot what you particularly want to see. You can also arrange to print out hardcopy of interesting plots.

Estimate which frequencies are present in the original signal. *After* you have done this, look in the M-file foursig.m to check your answer and see how the signal was actually created.

23.4 Fourier analysis of the tidal data

Having understood the basic operation of the M-file (fouran) you can now adapt it to cope with the tidal data. Edit the load statement to use the tidal data (times.mat and heights.mat). Then make corresponding alterations to the rest of the code. Note that the sensible units for the frequencies and periods in this case will involve hours rather than seconds. Have you identified which variable holds the 'signal' and which holds the time for the tidal data? In the original version of fouran.m the signal was in s and and the time in t. To look at the tidal data just type, for example,

```
>> clear
>> load times.mat
>> load heights.mat
>> whos
>> plot( ....)  etc.
```

The exercises can be accomplished using a combination of techniques. These include:

• making plots of the data over a variety of ranges if necessary;

- studying the Fourier analysis plots and the spectrum results themselves;
- making simple parametrisations of the dominant sinusoidal behaviour as in equation (23.1);
- using `spline` to interpolate subsets of data.

Exercises

23.1   Check out the time interval between measurements and the total time covered by the tidal measurements which are supplied.

23.2   Look at the maximum, minimum and mean of the data.

23.3   Plot the height vs time over different time scales – a day, a week etc.

23.4   Use the function `locmax` to estimate the heights of the high and low tides. Are these really the actual high tide values?

23.5   Estimate as best you can, the typical time between consecutive high tides.

23.6   Find on which days the tidal variation is strongest/weakest.

23.7   Find the dates covered by the given tidal data and the mean measured tidal height in this period.

23.8   Find the absolute maximum and minimum measured heights and when they were recorded.

23.9   Estimate the highest tide (whether measured directly or not) and when it occurred. Do the same for the lowest tide.

23.10   Find the dominant frequency components in descending order of importance. Comment on the numerical values of the corresponding periods.

23.11   Estimate the average time between consecutive low or high tides.

23.12   Estimate the sea level at 3.10 pm on 13 January and at at 6pm on 4 March in the year 1992.

23.13   Construct a plot of the rate in metres/second at which the tide is going out during the afternoon of 22nd January 1992 over a beach which has a gradient of 1 in 200 (assuming the tidal variation is independent of the local geography). What is the maximum speed achieved across the sand and at what time is it achieved? You should use the parametrisation (23.1) to help you do this.

Appendix 1
MATLAB Command Summary

Most commands can be found by using `help` but here is a quick and compact summary. For further details, type: `help` *command*.

Basic operations

`help` *fname*	displays help comments at the top of M-file *fname*
`quit`	quit MATLAB
`exit`	same as `quit`
`type` *fname*	lists contents of M-file *fname*
`who`	lists currently used variables
`whos`	as above but with more detail
`clear`	clear variables from memory
`what`	directory listing of all M-files on disk
`which`	locate M-files
`format`	change display format of results
`demo`	runs some MATLAB demos

Built in values

`pi`	π	`i,j`	$\sqrt{-1}$
`inf`	∞	`ans`	current answer
`flops`	floating point count	`clock`	wall clock time

Arithmetic and matrix operations

+	add numbers (scalars), vectors or matrices
–	as for add
*	multiplication of numbers or compatible matrices, vectors
.*	element-by-element multiplication of same size vectors, or matrices
/	division of numbers, right division of compatible matrices
./	element-by-element division of same size vectors, or matrices
\	left division of compatible matrices
^	power of number or square matrix
.^	element-by-element power of vector or matrix
'	transpose
size	size of matrix or vector
length	length of a vector
sum	sum of elements of vector
norm	magnitude of a vector

Common maths functions

sin, cos, tan	usual trigonometric functions
acos, asin	inverse of these
exp, log	exponential and natural logarithm
sqrt	square root
rand	random numbers in $[0, 1)$
round	round to nearest integer
fix	round towards zero
abs	absolute magnitude of real or complex number
angle	phase or argument
real, imag	real and imaginary parts
conj	complex conjugate

Matrix functions

det	determinant of square matrix
eig	eigenvalue and eigenvectors of a square matrix
inv	inverse
rref	reduced row echelon form
rank	rank of a matrix

Graph operations

plot	plot vector of points (linear X, Y-plot)
hold on/off	superimpose plots
clg	clear graphics screen
mesh	3D mesh surface
meshdom	domain for mesh plots
contour	contour plot
bar	bar graph
title	add a title
xlabel, ylabel	add labels
axis	axis scaling
text	annotate with text
print	print current graph

Function operations

fmin	minimum of function of one variable
fmins	minimum of function of several variables
fzero	find zero of function of one variable
spline	find spline function for data vector
quad	numerical integration
ode23	2nd/3rd order Runge–Kutta solution of ODEs

Statistics operations

mean	mean of data vector
std	standard deviation
cumsum	cumulative sum
cov	correlation between data vectors
min, max	minimum and maximum

Load/save operations

`save`	save variables to file
`load`	load variables from file
`diary` *fname*	log all MATLAB operations to file *fname*
`chdir`	change file directory
`dir`	list contents of directory

Appendix 2
Symbolic Calculations within MATLAB

MATLAB version 5 provides an interface calling Maple† commands directly for symbolic calculations. As this topic can be useful for readers, but is not essential for this book, we here use the same examples that appeared in this book for illustrations.

Differentiation and nonlinear solution — Chapter 4

To find $g = \frac{\partial f}{\partial t}$ for $f = x + 2ty - t - 2t^3$, and then solve for y from $g = 0$, do the following:

```
>> maple(' f := x + 2*t*y - t - 2*t^3 ;' )
>> maple(' g := diff(f,t) ')
>> maple(' h := solve(g,y) ')
```

Here the semicolon is only optional as MATLAB will add one to meet the Maple syntax. The output will be

```
f := x+2*t*y-t-2*t^3
g := 2*y-1-6*t^2
h := 1/2+3*t^2
```

Solution of differential equations — Chapter 7

To solve the ODE, $\frac{dx}{dt} = \frac{x}{t} - x$, try the following

```
>> maple(' a :=  diff(x(t),t)=x(t)/t - x(t) ')
>> maple(' b := dsolve(a,x(t)) ')
```

---

† Maple toolbox for MATLAB is developed and distributed by Waterloo Maple Software, Inc. See the web site: http://www.maplesoft.on.ca/ for details of ©Waterloo Maple.

The output will be (_C1 denotes a generic constant)

```
a := diff(x(t),t) = x(t)/t-x(t)
b := x(t) = t*exp(-t)*_C1
```

Integration — Chapter 11

To verify that $\int_{y_0}^{y_1} \frac{1}{\sqrt{c-y}} dy = 2\left(\sqrt{c-y_0} - \sqrt{c-y_1}\right)$ for $c = 2$, try

```
>>  maple( 'c := 2' )
>>  maple( 'a := int(1/sqrt(c-y), y=y_0..y_1) ')
```

The result will be

```
a := -2*(2-y_1)^(1/2) + 2*(2-y_0)^(1/2)
```

Solution of linear systems — Chapter 16

To check that the solution of

$$\begin{bmatrix} 1 & 1 & 0 & 3 \\ 2 & 1 & -1 & 1 \\ 3 & -1 & -1 & 2 \\ -1 & 2 & 3 & -1 \end{bmatrix} \begin{pmatrix} x_1 \\ x_2 \\ x_3 \\ x_4 \end{pmatrix} = \begin{pmatrix} 4 \\ 1 \\ -3 \\ 4 \end{pmatrix}$$

is $x = [-1\ 2\ 0\ 1]^\top$, try

```
>> maple(' A := matrix([[1,1,0,3],[2,1,-1,1],
                        [3,-1,-1,2],[-1,2,3,-1]])')
>> maple(' print(A) ')
>> maple(' x := ''x'' ')
>> maple(' b := array(1..4,[4, 1,-3,4])')
>> maple(' x := linsolve(A,b)' )
```

Note that the first two lines for defining A should be typed in as one line — here the 'breaking' is due to text processing only. The final result will be

```
x := VECTOR([-1, 2, 0, 1])
```

To get more help on Maple, type

```
>> mhelp          % Also maple('help')
>> mhelp solve    % Here for help on "solve"
```

Appendix 3
List of All M-files Supplied

The following lists all M-files which we provide for the reader via the internet from

  http://www.cup.cam.ac.uk/Scripts/webbook.asp?isbn=0521639204

or

  http://www.cup.cam.ac.uk/Scripts/webbook.asp?isbn=0521630789

These are listed under the chapters in which they are first used. Several are of course reused in subsequent chapters.

To locate the exact page number where an M-file or a MATLAB command is used in the book, check the index pages under headings 'M-files' and 'MATLAB commands'.

Part one: INTRODUCTION

Chapter 1

Chapter 2

- leslie.m

Chapter 3

- fibno.m
- hail.m
- gcdiv.m
- pow.m

Chapter 4

- tsine.m
- tsine2.m
- polyex.m
- goatgr.m
- goatfn.m
- parnorm.m
- linenv.m
- cubics.m
- hypocyc.m

Chapter 5

- tomato.m
- toms.m
- diabetic.m
- marks.m
- mannheim.m

Chapter 6

- c6exp.m
- exprand.m
- normrand.m
- randme.m
- unirand.m

Chapter 7

- fodesol.m
- fnxt.m
- odegr.m
- mode23.m
- species.m
- specfn.m
- ode23k.m
- vderpol.m
- vdplfn.m
- diffeqn.m
- dfeqfn.m

Part two: INVESTIGATIONS

Chapter 8

Chapter 9

- gcdran.m
- primes.m
- psp2.m
- miller.m

Chapter 10

- hypocy.m
- linenv.m
- paramc.m

Chapter 11

- zz1.m
- slide1.m
- slide1fn.m
- slide4.m
- slide4fn.m

Chapter 12

- mobius.m
- mobius1.m
- cobm.m
- cobq.m
- matit2.m
- matit3.m
- quadn.m
- perdoub.m

Chapter 13

- cnr1.m
- cnr2.m

Chapter 14

- randperm.m
- cycles.m
- riffle1.m
- riffle1a.m
- remm.m
- riffle1c.m
- ruffle1.m
- ruffle1c.m
- ruffle2.m

Chapter 15

- full_new.m
- gauss_ja.m
- f_rate.m
- cont4.m

Chapter 16

- lin_solv.m
- chop.m
- lu2.m
- lu3.m
- lu4.m
- solv6.m
- spar_ex.m

Chapter 17

- intdemo1.m
- intdemo2.m
- cont4.m
- cont7.m
- polyfit2.m
- polyval2.m

Chapter 18

See Chapter 7

Part three: MODELLING

Chapter 19

- queue.m
- exprand.m
- normrand.m
- unirand.m

Chapter 20

- fishy.m
- lmfish.m
- mparft.m
- mparst4.m
- resid4.m
- fishdat.m

Chapter 21

- fludat.m
- plagdat.m
- colddat.m
- sirepi.m
- sirfn.m
- mparst3.m
- resid3.m
- lagsum.m

Chapter 22

- topog.m
- snowsl.m
- snowboat.m
- snbtfn.m
- fsnow0.m
- fsnow1.m
- fsnow2.m
- snowmn.m
- snowmx.m

Chapter 23

- tides.m
- fourdat.m
- foursig.m
- locmax.m
- times.mat
- heights.mat
- noisyt.mat
- noisys.mat

Appendix 4
How to Get Solution M-files

Solution M-files and sketches of some sample solutions are available through the internet via a password made available to those using the book as a course text. Course leaders should contact the publisher – CUP:

for paperback

   http://www.cup.cam.ac.uk/Scripts/webbook.asp?isbn=0521639204

or for hardback

   http://www.cup.cam.ac.uk/Scripts/webbook.asp?isbn=0521630789

Appendix 5

Selected MATLAB Resources on the Internet

The name MATLAB stands for **mat**rix **lab**oratory. MATLAB was originally written to provide easy access to matrix software developed by the LINPACK and EISPACK projects, which together represent the state of the art in software for matrix computation. MATLAB converts user commands (including M-files) into C or C++ codes before calling other program modules (subroutines). Today MATLAB has been developed to cover many application areas.

To learn more about the product developers, check the web page

- http://www.mathworks.com/

If you hope to find out more of, or wish to use directly, the LAPACK and EISPACK programs, check the following

- http://www.netlib.org/lapack/

For users who would like a gentle and easy introduction to MATLAB, try the following

- http://www.ius.cs.cmu.edu/help/Math/vasc_help_matlab.html
- http://www.unm.edu/cirt/info/software/apps/matlab.html

For novice users who prefer short guides to MATLAB with concrete and simple examples, try

- http://www-math.cc.utexas.edu/math/Matlab/Matlab.html
- http://www.liv.ac.uk/pollo1/karen/document/486.dir/486.html
- http://www.indiana.edu/~statmath/smdoc/Matlab.html
- http://classes.cec.wustl.edu/~cs100/lab5/

Advanced users of MATLAB may check the following pages for more comprehensive guides to MATLAB

- http://wwwcache.rrz.uni-hamburg.de/RRZ/software/math/Matlab/
- ftp://ftp.math.ufl.edu/pub/matlab/

Even more sophisticated examples may be available on the internet; e.g. try the following for more M-files

- ftp://ftp.cc.tut.fi/pub/math/piche/numanal/
- ftp://ftp.mathworks.com/pub/mathworks/toolbox/matlab/sparfun/

The following sites contain technical information on MATLAB and FAQ (frequently asked questions) pages

- http://www.mathworks.com/digest/digest.html
- http://www.mathworks.com/newsletter/nn.html
- http://www.math.ufl.edu/help/matlab-faq.html
- http://ftp.mathworks.com/support/faq/faq.shtml
- http://www.uni-karlsruhe.de/~MATLAB/FAQ.html

References

[1]   H.Abelson and A.diSessa, *Turtle Geometry*, Cambridge, Mass: MIT Press, 1981.

[2]   J.W.Bruce, P.J.Giblin and P.J.Rippon, *Microcomputers and Mathematics*, Cambridge: Cambridge University Press, 1990.

[3]   P.Doucet and P.B.Sloep, *Mathematical Modelling in the Life Sciences*, Chichester: Ellis Horwood, 1992.

[4]   I.S.Duff, A.M.Erisman and J.K.Reid, *Direct Methods for Sparse Matrices*, Oxford: Clarendon Press, 1986.

[5]   G.Fischer, *Mathematical Models*, Braunschweig: Friedrich Vieweg Sohn, 1986 (Commentary volume, p.53.)

[6]   M.Gardner, *Mathematical Carnival*, Washington DC: Mathematical Association of America, 1989.

[7]   P.J.Giblin, *Primes and Programming*, Cambridge: Cambridge University Press, 1993.

[8]   L.M.Hall, 'Trochoids, roses, and thorns—beyond the spirograph', *College Mathematics Journal* **23** (1992), 20–35.

[9]   L.Haws and T.Kiser, 'Exploring the brachistochrone problem', *American Mathematical Monthly* **102** (1995), 328–336.

[10]   F.C.Hoppensteadt, *Mathematical Theories of Populations: Demographics, Genetics and Epidemics*, Philadelphia: Society for Industrial and Applied Mathematics, 1975.

[11]   J.F.Humphreys and M.Y.Prest, *Numbers, Groups and Codes*, Cambridge: Cambridge University Press, 1989.

[12]   N.Koblitz, *A Course in Number Theory and Cryptography*, New York: Springer-Verlag, 1987.

[13]   J.C.Lagarias, 'The $3x + 1$ problem and its generalizations', *American Mathematical Monthly* **92** (1985), 3–23.

[14]   H.-O.Peitgen and P.H.Richter, *The Beauty of Fractals : Images of Complex Dynamical Systems* Berlin: Springer-Verlag, 1986.

[15]   Y.Saad, *Iterative Solution for Sparse Linear Systems*, Boston: PWS Int. Thompson Pub. (ITP), 1996.

[16]   A.C.Thompson, 'Odd magic powers', *American Mathematical Monthly* **101** (1994), 339–342.

[17]   V.M.Tikhomirov, *Stories About Maxima and Minima*, Washington DC: Mathematical Association of America, 1990.

Index

301